A. Stapleton

Natural History of the Bible

A. Stapleton

Natural History of the Bible

ISBN/EAN: 9783743381964

Manufactured in Europe, USA, Canada, Australia, Japa

Cover: Foto ©berggeist007 / pixelio.de

Manufactured and distributed by brebook publishing software
(www.brebook.com)

A. Stapleton

Natural History of the Bible

Evangelical Normal Series.

Text-Book No. 9.

NATURAL HISTORY

—— OF THE ——

BIBLE,

Embracing a brief and concise description of the

TREES, PLANTS, FLOWERS, ANIMALS, BIRDS, INSECTS, REPTILES, FISHES and MINERALS MENTIONED IN THE BIBLE.

— BY —

REV. A. STAPLETON.

"The works of the Lord are great, sought out of all them that have pleasure therein."
Psa. 111. 2.

CLEVELAND, O.
Publishing House of the Evangelical Association,
LAUER & YOST, Agents.
1885.

AUTHOR'S PREFACE.

The preparation of this little work was undertaken at the earnest solicitation of friends. The author lays no claim to original investigations, and does not hope to add any new material to the natural history of the Bible. His only aim has been to present in a compact and popular form the researches of more eminent men. The writer hopes, however, that the perusal of this little volume will prove a source of pleasure and profit to the Bible student.

AMMON STAPLETON,

Seneca Falls, N. Y.

Natural History of the Bible.

THE STUDY OF NATURE.

The Creator has revealed himself to man, not only in the *Scriptures* of Divine truth, but also in *Nature*, for "The heavens declare the glory of God, and the firmament sheweth his handiwork, day unto day uttereth speech, and night unto night sheweth knowledge." (Psa. 19. 1, 2.)

It should be a pleasure to every Christian student, to study *both* volumes as from the hands of God. In no previous age has the study of Nature been pursued with such zeal as at the present day. There is absolutely nothing with which man is acquainted, that does not become an object of interest and profound investigation to the student of Nature. The pendulum of his thoughts constantly vibrates between the mighty orb, in the immeasurable distance of space, and the humblest worm that grovels in the dust. *Everything* that God has made is full of interest, and worthy of his profoundest investigation, for "The works of the Lord are great, sought out of all them that have pleasure therein." (Psa. 111. 2.)

The Creator himself first inducted man into the study of his works, and Adam was not only the first man, but also became, by Divine direction, the first naturalist, for we are informed that "Out of the ground the Lord God formed every beast of the field, and every fowl of the air, and brought them unto Adam, to see what he would call them, and whatsoever Adam called every living creature, that was the name thereof, and Adam gave names to all cattle, and to the fowl of the air, and to every beast of the field." (Gen. 2. 19.) In the writings of Moses, we find the earliest classification of natural objects extant, and so far as it goes, naturalists have not found it necessary to revise them materially, even in the present advanced state of science. The Scriptures abound with proof that its writers delighted in the study of the works of the Lord, and Solomon, at least, was an authority on natural history, for " He spake of *trees*, from the cedar tree, that is in Lebanan, even unto the hyssop that springeth out of the wall. He spake also of *beasts*, and of *fowls*, and of *creeping things*, and of *fishes.*" (1 Kings 4. 33.)

Since God has condescended to "clothe the lily," (Matt. 6. 28) and feed the young ravens when they cry, (Psa. 147. 9,) and watch every sparrow in its restless flight, (Matt. 10. 29,) should it be considered an unworthy subject for man to study the works of Nature with which he is surrounded, and then by faith look up through nature, to nature's God?

THE MINISTRY OF NATURE TO MAN.

Man is not alone in his desire to migrate to distant lands in quest of new homes and more congenial surroundings, sometimes he is preceeded, and sometimes followed by humbler forms of life, which the benevolence of the Deity has provided for his happiness and welfare. Although they write not the story of their migrations over sea and land, yet we know that they have made the long road delightful with fragrance and beauty, that leads back to "where Eden's bowers bloomed."

It is a well known fact that the most useful representatives of the animal and vegetable kingdoms have followed man in all his wanderings over the entire Globe. Impelled by some mysterious instinct, or directed by laws beyond the scrutiny of the most erudite naturalist, the choicest products of nature accompany and minister to him, who has Divine authority to subdue them, and make them subservient to his own welfare. (Gen. 1. 28, etc.)

Many of the products of nature have an interesting history. This is especially true of *flowers*. Some of our most beautiful varieties can be traced in their wanderings from America back to Europe, thence through many countries, and the lapse of many centuries, back to the distant Orient, and had we the facilities to do so, we could doubtless trace many a fragrant friend back to that primeval garden, where bliss and purity reigned supreme. When our first parents forfeited the favor of God, and were driven from the blissful bowers of Eden, it would seem as though every flower sympathized with their misery, every sparkling dew drop on the green leaf became a tear of pity,—every opening blossom became eloquent with prophecy,—"*forget-me-nots*" proclaimed that God had not "forgotten to be gracious." The white lily declared that purity should not perish from the face of the earth, while the blushing rose spake of goodness and beauty yet in store. Although compelled to grow in a soil cursed because of man's transgression, yet they refuse not to exhale their fragrance, neither do they hide their charms from the eyes of sinful man, but proclaim in language unmistakable, that "God is love." The direction of the Saviour to "consider the lilies," implies something more than a lesson on Christian resignation. When He says that "Solomon in all his glory was not arrayed like one of these," he proclaims the wisdom and skill which the great Creator bestows upon the humblest forms of life, as a constant memorial of his love to man.

Thus Nature ministers to the comfort and happiness of man. "*While the earth remaineth, seed-time and harvests shall not cease,*" (Gen. 8. 22,) but the book of Nature, as well as God's written word, shall continue to record the unspeakable benevolence of God to man. (Rom. 1. 19.) The birds that make the grove vocal with the melody of song, tell the Christian not to cease his singing until his voice is blended with the myriads on high, who sing the new song of redeeming love. (Rev. 5. 9.) The roar of the waterfall is to him a prophecy of that day of triumph when the saved shall mingle their voices "*As the voice of many waters,*" (Rev. 14. 2,) in praising a Saviour's love. When misfortune has nipped, not only the flowers that graced his garden, but perchance the blossoms of his heart, with confidence he says, "*The flower fadeth, but the word of our God shall stand forever.*" (Isa. 40. 8.) And when at last life's Summer-time is over, and the frosts of disease have nipped each bud and blossom of earthly prospect, he bows his head and mournfully says, with Isaiah, "*We all do fade as a leaf.*" (Isa. 64. 6.) The dark shadow of the death angel falls athwart his pathway, yet once more he cries, "*Bless the Lord, oh, my soul, and all that is within me, bless his holy name, who forgiveth all thine iniquities, who healeth all thy diseases, * * * so that thy youth is renewed like the eagles,*" (Psa. 103. 1–5,) and he takes his departure from earth, for the Summer-land above, where flowers no more shall fade.

PALESTINE, PAST AND PRESENT.

That Palestine was once a land "flowing with milk and honey," does not depend upon Scripture alone for confirmation. The observations of the traveler but confirm history, both sacred and profane, that it was once "a delightsome land." (Mal. 3. 12.) No country on the face of the globe, of equal area, has such great inequalities of surface as Palestine. Within the bounds of this little country, in one day's travel, may be experienced the heat of the Tropics, and the cold of the Arctic regions!

Mount Hermon lifts his snow-capped head 11,090 feet *above,* while the Dead Sea is *depressed* 1337 feet *below* the level of the Sea. The reader can hardly form an accurate idea of the great climatic differences arising from the great difference in levels. Perpetual Summer and perpetual snow is found in this land.

In early times, Palestine was a land of surpassing fruitfulness. The numerous hills, which are now bare and desolate, were once terraced, and covered with vines and olive trees, but after several thousand years of war and spoliation, between Barbarian and Jew, and Christian and Mohammedan for its possession, desolation has indeed overtaken this fair land. The terrace walls on the hill sides have been destroyed, and the rich soil has been washed into the valleys below, consequently the hillsides are no longer fruitful, but present a barren and desolate appearance.

Great changes have therefore occurred in the botany of Palestine. There is reason to believe that various trees and plants, which were abundant there when the Jews first gained possession of it, are now rarely met with, and in some instances extinct. This is also true of the Animal kingdom. Since the invention of fire-arms, the dangerous beasts of prey, such as the lion, leopard, wolf, bear, etc., have almost entirely disappeared.

During the last decade Palestine has received many tokens for good. Extensive explorations have been carried on under the direction of the British and German Governments. Large numbers of Jews have returned to the land of their fore-fathers, Christian colonies have been planted, notably at Jaffa (the ancient Joppa), and may we not hope that this garden of the Lord, which is the pivotal point in the history of the World, will be made to bloom again in its pristine glory, as a precursor of the World's millennium?

BOTANY.

There is, perhaps, no other country in the world that has a greater climatic range, in proportion to its area than Palestine. This gives it, necessarily, a very diversified flora. Palestine may be divided into three botanical regions :

First. The shore plain. The vegetation of this narrow strip of land is semi-tropical, and belongs to the flora of the Mediterranean basin.

Second. The lofty mountains and table-land, which comprise the greater part of the country. The flora of the uplands is that of more temperate countries, and comprises the oak, pine, cedar, ash, etc.

Third. The Jordan valley, which is about one hundred and fifty miles long, and from ten to twelve wide. Owing to the great depth of this valley, it has an almost tropical temperature. The vegetation of this valley is almost identical with that of Arabia and the valley of the Nile in Egypt.

ALGUM (2 Chron. 2. 8). Most writers are agreed as to the identity of the algum with what is now known as sandal-wood (*Santalum*). The sandal-wood is found in most Eastern countries, and consists of many varieties. The two varieties most valuable are the red sandal-wood (*Santalum Freycincti-anum*), and the white sandal-wood (*Santalum Album*). The wood forms a very important article of commerce. The red sandal-wood is supposed to be the algum of the Scriptures. It is a tree of medium size. Its wood is noted for its hardness, durability and fragrance, and is of a beautiful garnet color when finished. Hiram, king of Tyre, furnished a quantity of this valuable timber to be used in the construction of the temple.

It would seem from 1 Kings 10. 11, 12, and 2 Chron. 9. 10, 11, that the algum and almug are identical.

ALMOND (*Amygdalus communis*). The almond is frequently mentioned in the Bible, and is known to have been very abundant in Palestine. It belongs to the same genus as the peach, which it closely resembles, except in its fruitage. The tree, which grows in nearly all warm countries, attains a height of twenty-five to thirty feet. The almond fruit grows in a sort of shell or husk, which opens of its own accord as the fruit ripens. The almond nut is the kernel of the fruit. The tree blooms very early in the Spring, putting forth its beautiful white flowers in great profusion long ere the leaves make their appearance.

ALMOND.

ALOES (*Aquilaria agallochum*). Also called lign aloes (Num. 24. 6). This tree, of which there are many varieties, is very abundant in Asia. The aloes of Scripture must not be confounded with the bitter and nauseous drug of that name, which is the product of a smaller variety, nor with the American aloe (*agave Americani*), which is an entirely different thing. The aloes of Scripture is a medium-sized tree, with large, spreading branches, and makes a very fine appearance. It is chiefly valuable for the resinous substance or gum obtained from it. This product of the aloe was much prized by the ancients for its delightful fragrance (Sol. Song 4. 14). It also possesses great preservative properties, and was used in embalming the dead (John 19. 39).

Lign aloes (Num. 24. 6) is a corruption of lignum aloes (aloes wood).

ANISE. It is universally believed that the marginal rendering (*dill*) is the correct one (Matt. 23. 23). The word translated anise in our authorized version is rendered "dill" in nearly all other European translations.

DILL (*Anethum graveolens*) is common to the warm countries bordering on the Mediterranean Sea. It is closely allied to our fennel. The seeds of the dill, which have an agreeable, aromatic taste, are chiefly used for flavoring, and as a condiment in food.

APPLE (Sol. Song 2. 3–5; 7. 8). Much difference of opinion exists as to the true meaning of the Hebrew word "Tapuach," rendered "apple" in the authorized version. A few writers maintain the correctness of the translation, while the greater number say that the citron (*Citrus medica*) is meant. Good authorities hold that the apple was introduced into Palestine at a much later period, while the citron tree has been cultivated there from remote antiquity. The citron tree is of the same genus as the orange, which it resembles. The fruit, however, differs very much, being much larger and having a ribbed and warted surface. Although somewhat acid, the fruit is fragrant and pleasant to the taste.

ASH. "He planteth an ash" (Isa. 44. 14). Considerable uncertainty exists as to the true meaning of the Hebrew word "oren," which is rendered pine in the Vulgate and Septuagint, cedar in the German and Portuguese, and ash in the English and other European versions. There are no valid reasons, however, for doubting the correctness of our version, as the ash is known to be native to Palestine and other Asiatic countries. There are many varieties of ash, the wood of which is very tough, flexible and durable.

The principal American varieties are the white ash (*Fraxus Americani*), the black ash (*F. pubescens*), and the water ash (*F. Sambucifolia*).

BALM (Gen. 37. 25). This is supposed to have been the resinous substance of the balm of Gilead tree (*Balsamodendron Gileadense*). It is a native of Abyssinia, but is also found on the mountains of Gilead, and hence the name. The resin or gum of this tree is valuable chiefly on account of its healing properties.

"Balm of Gilead" is an entirely different tree from that

known in America by that name, which is the Balm-of-Gilead fir (*Abies Balsamea*).

BARLEY (*Hordeum*). A very important and extensively cultivated cereal, mentioned quite frequently in the Scriptures. In Palestine, barley is sown both in the Spring and Autumn. The later sowing forms the principal crop, which is gathered in March and April, and is known as the "barley harvest" (Ruth 1. 22). Barley was fed to cattle (1 Kings 4. 28), and was made into bread (Judg. 7. 13; John 6. 9). Barley is one of the hardiest cereals known, growing not only in semi-tropical countries, but also in very cold regions. It is said to produce good crops in the Himalaya mountains, at an elevation of fourteen thousand feet above the level of the sea.

BAY TREE. This tree is mentioned but once in the Bible (Psa. 37. 35): "I have seen the wicked in great power, spreading himself like a green bay tree." The "bay" tree is recognized as the laurel (*Laurus nobilus*), and is so rendered in Luther's version (*Lorbeerbaum*). This species of laurel is now very rare in Palestine. It is a much larger tree than our American varieties. Having a bushy and very pretentious appearance, the comparison in the above passage is obvious. It is from the stiff, green leaves of this tree that crowns were made for the victors in the Olympian games.

BEAN (*Faba vulgaris*) occurs twice in the Bible (2 Sam. 17. 28; Ezek. 4. 9). The bean has been cultivated from time immemorial in the East, and furnishes a wholesome farinaceous article of diet. There are very many varieties of this plant, and a description of them seems unnecessary.

BOX (*Buxus sempervirens*). The name of this shrub or tree occurs in Isa. 41. 19, and 60. 13. In this country it is a mere ornamental shrub, while in Southern Europe and Asia it attains a height of from twenty to twenty-five feet. It is an evergreen, and remarkable for the closeness of its leaves and branches, which form a very compact mass. Boxwood, with one or two exceptions, is the hardest and heaviest wood known, being too heavy to float on water. It is the best wood known for engraving and carving purposes. It is also much used in the manufacture of musical and mathematical instruments.

BRIERS occurs quite frequently in the Bible, and is a general term to denote any thorny or prickly plant. (See *Thorns and Thistles.*)

BULRUSH (Ex. 2. 3); also called rushes (Job 8. 11; Isa. 9. 14). This is, without doubt, the famous papyrus of the ancients (*Cyperus papyrus*). It is a species of sedge, growing in the shallow margins of rivers. It has remarkably large and extensive roots in proportion to its size, which ranges from eight to ten feet. The stem is soft and cellular, triangular in shape, and tapering from the bottom upward. The plant is terminated by a calyx containing numerous pendant spicules. The papyrus or paper was made by laying the stems in a row, and then laying another row crosswise, and subjecting it to great pressure. The product was a paper almost as durable as parchment. Specimens have been taken from tombs and ruins several thousand years old.

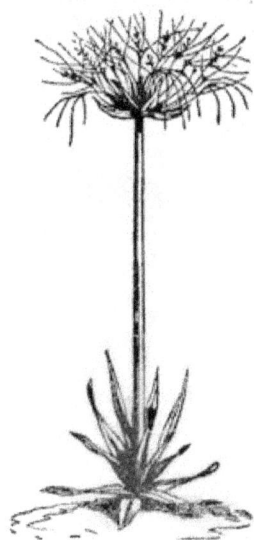

BULRUSH.

CALAMUS (Sol. Song 4. 14; Ezek. 27. 19), or Sweet Calamus (Ex. 30. 23), Sweet Cane (Isa. 43. 24; Jer. 6. 20), "were all probably the same plants, or at least belonged to the same genus. It was produced in Arabia and India, and of an inferior quality in Egypt and Syria. It was one of the ingredients of the sacred ointment, and an article of Syrian commerce. It grows about two feet in height, is very fragrant, and resembles common cane." (*J., F. and Brown.*)

CAMPHIRE. "My Beloved is as camphire" (Cant. 1. 14; also 4. 13). This plant or bush must not be confounded with camphor, which is derived from an entirely different plant (*camphora officinarum*). The flowers of the camphire grow in clusters, like the lilac, and are very beautiful. From the pulverized leaves of the camphire the ancients manufactured a dye of a beautiful orange color, with which the females formerly stained their lips and finger nails.

CEDAR OF LEBANON.

CASSIA.   "Thy garments smell of cassia" (Psa. 45. 8; also
Ex. 30. 24; Ezek. 27. 19).   Supposed to be the product of
Aucklandia costus, a plant growing chiefly in India.   It is a
delightful perfume, and was used for that purpose by the ancients.

CEDAR (*Cedrus libani*).   There are many and widely-distributed varieties of cedar, but that known to botanists as
Cedrus libani is of the greatest interest to Bible students.
The "cedar of Lebanon" is now almost extinct, but was very
plentiful on the mountains of Palestine in ancient times.   It is
not, as we might suppose, a tall tree, but attains a remarkable
thickness.   Of the few still standing on Mount Lebanon, there
is one sixty-three feet in circumference.   The branches are
very numerous and extended.   On account of the durability of
the wood, it was used in the construction of the temple (1 Kings
5. 6); also used for ship-building (Ezek. 27. 5).   Because it
was evergreen, strong, durable and beautiful, it was used as an
emblem of the righteous (Psa. 92. 12).

CHESTNUT (*Castanea*) Gen. 30. 37; Ezek. 31. 8).   Supposed
by Kitto and others to be the Plane tree (*Plantanus orientalis*),
which is the rendering in both the Vulgate and Septuagint.
Good authorities, however, maintain the correctness of our version, as it is known to the native in the East since it derived
its name from Castaneum, a town in Asia Minor.   The chestnut has long, smooth, serrated leaves.   Most species of chestnut produce nuts, which are used as food.   The chestnut
attains an enormous size.   One on Mount Etna was two
hundred and four feet in circumference.   Its wood is strong
and durable, and much valued for manufacturing purposes.

CINNAMON (*Laurus cinnamonium*).   A tree native to nearly
all tropical countries.   The tree attains a height of from twenty
to thirty feet, and about a foot in thickness.   Leaves oval, from
four to six inches in length.   The flowers of the cinnamon are
gray externally, and pale yellow internally.   It bears a fruit
resembling an acorn.   The dried bark of the small shoots and
branches form the cinnamon of commerce.   It is very fragrant
and aromatic (Ex. 30. 23; Prov. 7. 17; Cant. 4. 14).

COCKLE (Job 31. 40).   Hebrew, Besha.   Literally, noxious
weeds.   There is no agreement whatever among authors as to

the meaning of this term. Luther has it "Dornen"—thorns. Some suppose it was the Aconite plant (*Aconitum album*), a very poisonous plant. Others refer it to the nightshade (*Solanum*), a poisonous vine.

CINNAMON.

CORIANDER (*Coriandrum sativum*) (Ex. 16. 31; Num. 11. 7). A small, annual plant, native to Southern Europe and the East. It is from one to two feet high, and has bipinnate leaves. It produces an aromatic seed used in medicine; also for flavoring.

CORN occurs frequently in the Bible. It is a general term,

2

and comprised all kinds of grain. Our "corn" (*maize*) is a native of America, and was unknown to the ancients.

CUCUMBER (*Cucumis sativus*). A vine cultivated from the earliest times for its fruit. It grows in all semi-tropical and temperate climates. Although the fruit is considered by us unwholsome, it was much esteemed by the ancients (Num. 11. 5; Isa. 1. 8).

CORIANDER.

CUMMIN (*Cuminum cyminum*). An annual plant native to Egypt and neighboring countries. Its stem is branched, and has thread-like leaves. The flowers are small, and of a white or pink color. The seeds have a warm and agreeable flavor, and are used as a condiment and medicine (Isa. 28. 25; Matt. 23. 23.)

CYPRESS (Isa. 44. 14). This tree is believed to be the species of cypress known in botany as *Cypressus Sempervirens.* It is a native of Asia Minor, Northern Africa and Southern Europe. It is a tree of medium height, tapering in form, quadrangular twigs and dark-green leaves, and, therefore, since the earliest times, an emblem of mourning. The wood of the cypress is of a reddish hue, very hard and compact, and takes a fine finish. It is thought to be the most durable wood known. There are specimens in European museums known to be several thousand years old.

DOVE'S DUNG. This term occurs only in 2 Kings 6. 25, where it is said that, in the famine at Samaria, "a fourth part of a cab (nearly a pint) of dove's dung was sold for five pieces of silver." It is generally agreed that the bulbous root of a small plant indigenous to that country is meant. It is called "Star of Bethlehem," and is known in botany as *Onithogalum umbellatum.* The bulb may be eaten raw, or dried, pulverized and mixed with flour, and baked.

EBONY (Ezek. 27. 15) occurs but once in Scripture. Ebony is the heart wood of a tree known in botany as *Diospyros Ebenum.* There are also several less valued varieties. The tree, which is found chiefly in Ceylon and India, grows to an enormous size. The ebony of commerce is a jet-black wood, extremely hard and susceptible of a high polish.

FITCH (Isa. 28. 25; Ezek. 4. 9). A species of pea known to botany as *Vicia Sativa.* The "fitch," or "vetch," as we now call it, grows in all temperate latitudes. In England it is much used as a green fodder for cattle. This species of pea is now seldom used as human food.

FIG. The name of a tree most frequently mentioned in Scripture, the leaves of which constituted the first garments worn by mankind. The fig (*Ficus Carica*) is native to all warm climates, and is, perhaps, the most extensively cultivated tree in the world. There are over one hundred species, some of which attain a large size. That species best known to us is a low, deciduous tree, or shrub, with large, deeply-lobed leaves, which are rough on the upper and downy on the lower surface. The fruit, which is pear-shaped, is grown in the axiles of the leaves. The fruit has always been highly

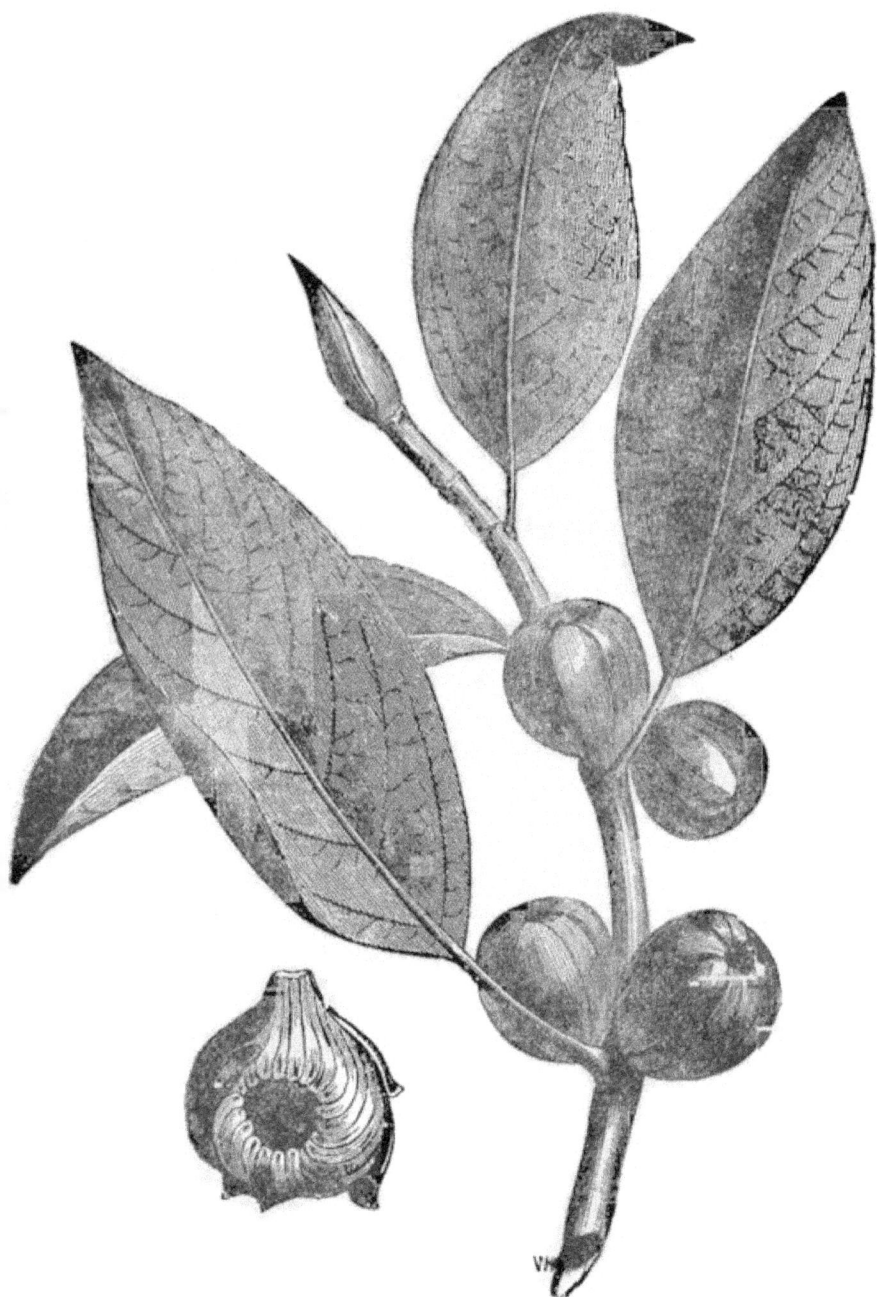

esteemed as food, and is a staple article of commerce in the East. The Hebrew name for fig—tamar—was used as a proper name for females (Gen. 36. 6; Ruth 4. 12: 2 Sam. 13. 1; Ezek. 47. 19).

FIR. A tree mentioned quite often in the Bible, and believed by the best authorities to be identical with the cypress mentioned in Isa. 44. 14 (*Cupressus Sempervirens*). A beautiful evergreen, allied to the pine; still found in the East; its wood is hard and very lasting, and much used for building purposes. (See Cypress.)

FLAG. The Hebrew word *achu*, rendered "meadow" in Gen. 41. 2, is rendered "flag" in Job 8. 11. Here it doubtless refers to some specific plant. "Can the flag grow without water?" From these passages it seems that "achu" refers to some water-weed; also used as pasture. The best authorities agree that the species of flag known in botany as *Cyperus Esculentus* is meant. This water-plant, in addition to its use as pasturage, has also an edible root.

FLAX. (*Linum Usitatissimum*). A small annual plant, mentioned very frequently in the Bible. It is found in nearly all temperate quarters of the world. It has a slender, erect stem, from two to three feet high, and branching only near the top. The leaves are lanceolate and small. The flowers are a beautiful blue. It is one of the most useful plants known to man, and has been cultivated from the earliest times. From the fibers of its inner bark linen is made, while the seeds, expressed, produce linseed oil. The residuum is oil-cake, so valuable for cattle; or, when ground, it forms oil-meal.

FRANKINCENSE (Ex. 30. 34). A very odoriferous resin which was mixed with other fragrant substances and burnt as a sacrificial offering by the Jews. It is not known definitely of what tree this gum is the product. Several good authorities suppose it was the *Boswelia serrata*—a tree still abundant in the Orient.

GALBANUM. The resin or gum of a plant known as *Bubon Galbanum*, still native to Syria. It formed an ingredient in the Holy perfume (Ex. 30. 34). This gum is still found in drug-stores, and is considered a valuable medicine.

GARLIC (*Allium Ascalonicum*). A bulbous plant found in

the East, and introduced into Europe by the Crusaders. As
there are many varieties of garlic, it is probable that the
Egyptian variety was more desirable as an article of food than
that known to us, or they would never have preferred bondage
under Pharaoh, with "the leeks, and the onions, and the
garlic," to liberty and the prospect of a rich country under the
government of God (Num. 11. 5).

GOPHER-WOOD (Gen. 6. 14). The name of the wood used in
the construction of the ark. It seems a hopeless task to determine definitely what wood is meant. The Septuagint rendering is "squared timbers;" the Vulgate, "planed wood;" the
Chaldee and others have it "cedar;" and Luther has it Fir,
"Tannenholz." Our Version does not venture a translation
of the Hebrew word gopher. The general supposition is, however, that the fir or cypress is intended.

GOURD. This word occurs twice in our version. In the
original, however, the words are different, and evidently refer
to different things. We are told in 2 Kings 4. 49, that "one
went out into the field to gather herbs, and found a wild vine,
and gathered thereof wild gourds his lap full," which proved
to be the "death in the pot." From the narrative it is quite
probable that it was the *Cucumis Prophetarum,* a very bitter
and poisonous vine, still abundant in Palestine.

Jonah's gourd (Jonah 4. 6–10), it would seem, was a different growth, and writers have never been agreed as to its
identity. It is maintained by some writers that a vine or
climbing plant is intended, while critical authorities give the
castor-oil plant (*Ricinus Communis*) as the proper plant, for
the following reasons. The Hebrew word *Kikayon,* rendered
"gourd" in the above passage, is preserved in form by Pliny,
who called the plant *Kiki,* and describes it as the castor plant.
In addition to this, the Arabic version makes the Ricinus the
gourd of Jonah, for the very good reason that the Arabic name
of the plant is of the same form as that of thn Hebrew, translated "gourd" in Jonah. The castor plant answers to the
description in Jonah very well. It is of rapid growth, attain-
ing a height of eight to ten feet, leaves sometimes a foot wide,
thus affording a grateful shade.

GRAPE (vine) (*Vitis Vinifera*). A deciduous climbing plant frequently mentioned in Scripture. The leaves of the grape vine are lobed, serrated, and more or less hairy. The stem sends out numerous branches, which attain a great length. The grape vine is found in all mild climates. In some countries it attains an enormous size. Sometimes the stem is eighteen inches in diameter. The large varieties are said to bear fruit for three hundred years. The expressed juice of the fruit forms wine. The fruit dried in the sun is known as raisins. There are said to be more than fifteen hundred varieties of grape.

HEATH occurs twice in Scripture, both times in Jeremiah, although dissimilar in the original. In Jer. 17: 6 it is supposed the true heath or heather is intended. The heath (*Erica Vulgaris*) is a genus of small shrubs, found not only in the Orient, but growing abundantly in Europe. The leaves are small, linear, and evergreen. It bears a calyx of four leaves, and ovate corolla. The flowers are spike-like, and of lilac rose color. The heath does not grow in America. Jer. 48: 6 is supposed to refer to the tamarix or juniper.

HAZEL (*Corylus*). A small tree or shrub, of which there are many varieties, one of which produces the nuts known as "filberts." The charcoal of hazel wood is much esteemed by artists for crayons. "Hazel" occurs in Gen. 30 : 37, when Jacob used rods of this tree for a peculiar purpose.

HEMLOCK (*Conium Maculatum*). The Hebrew word *rosh*, translated hemlock in Hosea 10: 4 and Amos 6: 12, is in other passages rendered gall and bitterness. Hemlock is a small and poisonous plant found in all Eastern countries. It has a root somewhat resembling a small parsnip. The stem is round, smooth and hollow, and of a green color. Its leaves are large and tripinnate. Extract of hemlock is a deadly poison.

HUSKS. The prodigal "would fain have filled his belly with the husks that the swine did eat" (Luke 15: 16). It is generally agreed that reference is here made to the pods of the carob-tree (*Ceratonia Siliqua*). The carob is a leguminous or pod-bearing tree, growing abundantly in Asia Minor and southern Europe. The tree is of medium size. Its leaves are pinnate, dark-colored and evergreen. The pod is from four to eight inches in length. The seeds are bitter and of no use, but the interior of the pod, or husk, is used as food by the poor.

HYSSOP (*Hyssopus Officinalis*). A small, shrubby plant — the upper part of the stems quadrangular—lanceolated, evergreen leaves —flowers of a bluish color. The hyssop was the symbol of purification by the law, and when our Saviour received it (John 19. 29), it was the last legal purification which the hyssop should make on earth. Henceforth, His blood should be the all-atoning sacrifice for sin, and the cross the symbol of its power.

JUNIPER (*Juniperus*). A genus of small trees, closely allied to the cedar. There are many varieties of juniper. It is an evergreen, and bears small berries of a blue-black color. which require two years to ripen. The berries of some varieties are edible. Extract of juiper is a valuable medicine (1 Kings 19. 4; Job. 30. 4; Psa. 120. 4).

KARPAS=COTTON. Karpas is a Hebrew word, translated green in Esth. 1. 6. It is said there "were white, GREEN (*Karpas*), and blue hangings." Critics are generally agreed that "Karpas" has reference to a cotton fabric, as the word is similar in meaning to "cotton" in the Sanscrit (*Karpasum*) Arabic (*Karphas*), and Persian (*Kirbas*). (See Kitto, E'cy.)

LEEK (*Allium Porum*). A biennial plant of the same genus as the onion, from which it differs in having thinner and longer bulbs. Its leaves, too, are wider than those of the onion. It has been cultivated for food from remote times, and was one of the five good things of Egypt for which the Israelites hungered in the wilderness (Num. 11. 5).

HYSSOP.

LENTILES. A small, leguminous plant, native to countries

bordering on the Mediterranean Sea. The stem is weak and branching, from ten to fifteen inches high. Leaves pinnate, flowers small, white, pale blue, or lilac color; pods short, thin and smooth, and contain two seeds. The lentile is much esteemed as food for man or beast. The red lentile probably constituted the "red pottage" for which Esau sold his birthright (Gen. 25. 30).

JUNIPER.

Lign Aloes *Aquilaria Agallochum*). The term "lign aloes" occurs in Num. 24. 6, and is the same plant already described. Lign is merely a corruption of lignum, the Latin for wood.

Lily. " Consider the lilies " Matt. 6. 26. In the Old Testament several different words have been rendered lily. There is, consequently, some doubt as to its correctness in some cases. In our Saviour's eloquent reference to it, the original is krinon (lily). Of the lily there are many species. The white lily (*Lilium Candidum*) is doubtless the one referred to by our Saviour. It is native to Palestine, and still found growing abundantly there. Its flowers are very large, pure white, and very fragrant.

MALLOW (Job. 30. 4). A species of herbaceous plants of the natural order *Malvaceæ.* It is a perennial plant, stem erect, flowers bluish-red and quite large. There is a podded mallow (*Hibiscus Esculentus Lin*), a native of the Levant. The mallow is sometimes used as food by the poorer classes,

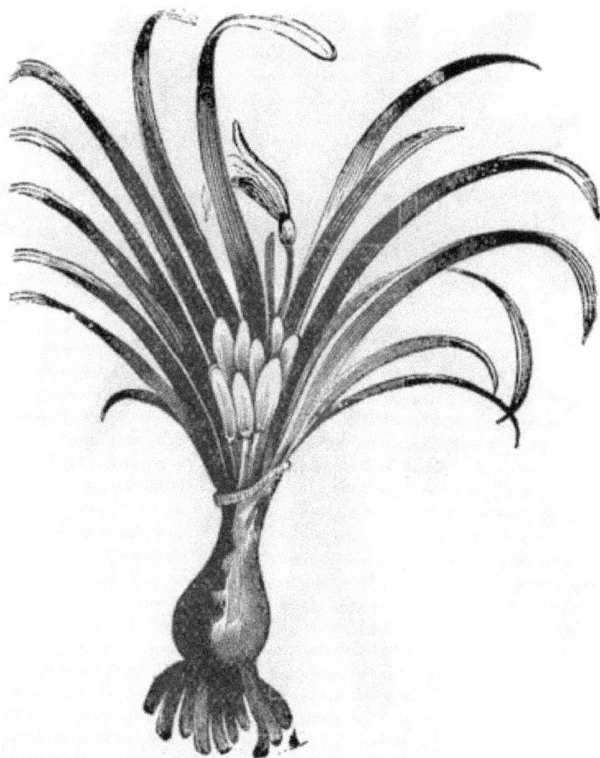

LILY.

MANDRAKE (*Atropa Mandragora*) (Gen. 30. 14; Cant. 7. 13). There are two varieties of mandrake, the vernal and autumnal. Both are natives to the East and South Europe. The root is carrot-like and quite large. The leaves, which are oblong-ovate, spring from the root. The berries sometimes attain the size of a small apple, and were formerly eaten. The mandrake possesses very strong medical (narcotic) properties.

MELON (*Cucumis Melo*), one of the five good things of Egypt which the Jews wishfully remembered in the wilderness (Num. 11. 5). The melon belongs to the same genus as the cucumber,

MANDRAKE.

and is found in all warm climates. It is an annual plant, long, trailing stem, lateral tendrils, rounded, angular leaves, flower small and yellow, fruit round or ovate. There are many varieties of the melon, some of which produce very delicious fruit.

MILLET (*Panicum Miliaceum*). Referred to in Ezek. 4. 9, where it is enumerated as one of the components of that bread which was a type of the nature of the prophecy. It is one of the common grains of Palestine, and has been cultivated there from the earliest times. There are numerous varieties of millet. Is is an annual grass, much valued as fodder. The grain is used in Asia as food.

MINT (*Mentha Viridis*). Mentioned by the Saviour (Matt. 23. 23; Luke 11. 42). Mint is a native of almost all the temperate parts of the globe. It has erect, smooth stems, from one to two feet high. The leaves are lanceolate, smooth and serrated. It has an aromatic, agreeable odor. Used as a medicine and condiment.

MULBERRY (1 Chron. 14. 14, 15; 2 Sam. 5. 23, 24). Considerable difference of opinion exists as to the Hebrew word baca, in the above passages rendered "mulberry." The eminent critic Rosenmüller prefers " pear-tree." The uncertainty arises from the supposed inappropriateness of the mulberry-tree in its connection with the circumstances narrated in the above passages. Several authors give the poplar as the tree designated. We may observe, however, that nearly all European versions give the equivalent of "mulberry." The mulberry (*Morus Nigra*) is once mentioned in the New Testament (Luke 17. 6), where the original is *Sukaminos*—the Greek for mulberry—rendered "sycamine." The mulberry is a native of the East, and was introduced into Europe over a thousand years ago. The common mulberry is a low tree, with numerous branches, rough bark, heart-shaped, and unequally serrated leaves. The fruit is excellent, while the leaves furnish food for silk worms. The red mulberry (*Morus rubra*) is a native of America. It is larger, and the timber more valuable than that of the Eastern varieties.

MUSTARD (Matt. 13. 31; 17. 20; Mark 4. 31: Luke 13. 19). Some difficulty has been felt by writers in understanding the assertion of the Saviour, in the parable of the mustard, that "it grew, and waxed a great tree, and the fowls of the air lodged in the branches of it" (Luke 13. 19). The eminent Oriental botanist, Prof. J. F. Royl, holds that it is not the mustard plant (*Sinapis nigra*), but a tree known in botany as *Salvadora Persica*. It formerly grew abundantly on the shores of the Sea of Galilee. The plant has a very small seed, which produces a

large tree with numerous branches. The seed has the same
properties, and is used for the same purposes as mustard. The
weight of authority, however, inclines to the authorized version,
and there need be no difficulty in receiving the Saviour's
description of the mustard plant literally, when it is remem-

MYRRH.

bered that there are several varieties of the mustard in Syria
and Palestine, one of which, found on the plain of Esdraelon,
grows to the height of from nine to ten feet, producing branches
sufficiently large to bear the weight of a bird.

MYRRH (*Balsamodendron Myrrha*). A tree found most
abundant in Arabia. It is rather small and shrubby, bark
greyish white, leaves obovate and serrated. The Scripture pas-

sages below relate to the gum which is obtained by incision. It
was fragrant (Sol. Song 4. 6); presented as a gift to the infant
Saviour (Matt. 2. 11); mingled with the drink offered to him on
the cross (Mark 15. 23); used to embalm him (John 19. 39).

MYRTLE (*Myrtus Communis*). A shrub or small tree, found
abundantly in the East. The tree is evergreen. The leaves,
which are opposite, are ovate or lanceolate, and of a dark-green
color. It bears a beautiful white flower. It makes a fine
appearance, and is much esteemed (Isa. 41. 19; 55. 13; Neh.
8. 15; Zech. 1. 8).

NARD=SPIKENARD. An ingredient of the ointment with
which Mary anointed the Saviour (Mark. 14. 3; John 12. 3).
Eminent authorities, such as Sir William Jones and Prof. J. F.
Royl, agree that nard or spikenard was the product of a plant
known to botany as *Nardostachys Jatamansi*. There are two
species of the nard, one of which is found in the Alpine regions
of Europe, and the other a native of India. The Indian spec-
imen is that called *N. Jatamansi*. It rises from the ground
like a hairy spike of bearded wheat. The root, which is from
several inches to a foot in length, sends up many stems with
purple flowers. The root possesses a very delicate odor.

OAK (*Quercus*) (Gen. 35. 8; Isa. 2. 13; 6. 13; Hos. 4. 13;
Zech. 11. 2). The oak is one of the most widely-distributed
trees on the globe. There are very many species of the genus
Quercus, some of which have hardly any resemblance to each
other. The most valuable varieties grow in the temperate zone.
It is likely that the oaks of Scripture are the evergreen varieties,
which are found in the East. They are the *Q. Ilex*, *Q. Gra-
muntia*, and *Q. Coccifera*. The acorns of some species are
edible. From its size, durability and stateliness, the oak well
deserves the title, "King of the Forest," often given it. Several species sometimes attain a height of one hundred and
eighty feet; and some, serving as landmarks in Europe, are
known to be a thousand years old.

OIL-TREE (Isa. 41. 19). The olive-tree.

OLIVE (*Olea Europœa*). The name of a tree found abundantly in the East, and now naturalized in all warm countries
in Europe. The olive is an evergreen. In the cultivated

varieties, the leaves resemble those of the willow, are lanceolate, dark-green color above, and whitish-grey beneath. Flowers
white. Fruit about the size of pigeons' eggs, and generally
green. The olive is an enormously fruitful tree, and attains a
great age. Mentioned twenty times in Scripture. Used alle-
gorically (Judg. 9. 8); architecture (1 Kings 6. 23); emblem of
beauty (Hos. 14. 6); of fatness (Rom. 11. 17).

OLIVE.

ONION (*Allium Cepa*). One of the good things of Egypt for
which the Israelites pined during their journey in the wilderness (Num. 11. 5). The onion is a biennial, bulbous plant,
with a swelling stem, leafy at the base. Tapering, fistular
leaves. The onion has been cultivated since remote antiquity.

In Egypt it grows much larger, and is less acrid and better-flavored than what is grown here.

PALM. A genus of plants which comprises over five hundred species, and are found in every tropical and semi-tropical quarter of the globe. The only species referred to in the Bible is the date palm (*Phœnix Dactylifera*). This species, which is the most important and valuable of all, is found in the depressions of Palestine. The tree is slender, branchless, and very flexible, and attains a height of from thirty to sixty feet. It bears a head of from forty to eighty glaucous, pinnated leaves of enormous size. The fruit, which is known to us as dates, grow in clusters that weigh from twenty to twenty-five pounds. In a general sense, the palm is the most valuable tree known to man ; every part of it is utilized. It is the chief wealth and sustenance of millions of people in Asia and Africa. The fruit is the most nutritious food known, containing fifty-eight per cent. of sugar. The tree is mentioned twenty times in Scripture. Because of its uprightness, flexibility, fruitfulness, and evergreen character, it is a befitting emblem of the Christian (Psa. 92. 12).

PINE (Isa. 41. 19 ; 60. 13). Bible botanists consider it somewhat doubtful whether the pine is intended in the above passages. Although several species of pine are found in Palestine, it is held that the Hebrew word so rendered is incorrect. The eminent Hebrew scholar, Gesenius, in his critical works, makes it the holm-tree, which is a species of holly (*Ilex*). This rendering is followed by most modern commentators. The holm is an evergreen, and answers the references in the above passages very well.

POMEGRANATE (*Punica Granatum*). Of all the trees or plants thus far described, none are more beautiful or important than the pomegranate. It is frequently mentioned in the Scriptures, and also in classical literature. Many of the Greek deities with Jupiter, Juno and Venus, are often represented as holding a pomegranate. It can hardly be called a tree, as it seldom reaches the height of twelve feet. The leaves are large and of a dark-green color. Its fruit is about the size of an orange, with a thick leathery rind of a fine golden yellow, with a rosy tinge on one side. The fruit is pulpy and full of reddish seeds, which are also edible. The pomegranate is of a delicate flavor, and much esteemed. A delicious drink is made from the juice of the fruit (Sol. Song 8. 2). The whole plant possesses medicinal properties, while the finest morocco leather in the world is tanned with a preparation of pomegranate bark.

POPLAR occurs in Gen. 30. 37 and Hosea 4. 13, and is supposed to be the white poplar (*Populus Alba*), which grows plentifully in the East. The poplar is of very rapid growth, and attains an immense size. Its leaves are heart-shaped or lobed, and serrated — smooth, glossy and dark-green above, and silvery-white and downy beneath. Its wood is soft and much used in cabinet-making.

PULSE occurs in 2 Sam. 17. 28 and Dan. 1. 12, 16. Pulse is the edible seeds of pod-bearing plants, such as beans, peas, etc. Pulse, in its various forms, has always been an important article of food in the East.

POMEGRANATE.

REED. Several different words are translated reed in the authorized version. The Hebrew word *Kaneh*, rendered "reed" in 1 Kings 14. 15; 2 Kings 18. 21; Job 40. 21; Isa. 19. 6, etc., doubtless refers to the *Arundo Donax* of botany. This reed is slight and slender, and yet very strong, and was much used by the ancients for arrow-shafts. Its principal use, however, was for writing purposes. The reed was first soaked in water containing certain ingredients. It was then carefully dried. By this process it was hardened, while the pith was almost absorbed. Its excellency was hardly excelled by the modern quill. In the New Tes-

3

tament, this plant is embraced in the Greek word *Kalamos,* which is a somewhat general term. *Kalamos* is rendered "reed" in Matt. 11. 7; Luke 7. 24; "pen," in 3d John 13: "I will not write with pen (*Kalamos*) . . . unto thee."

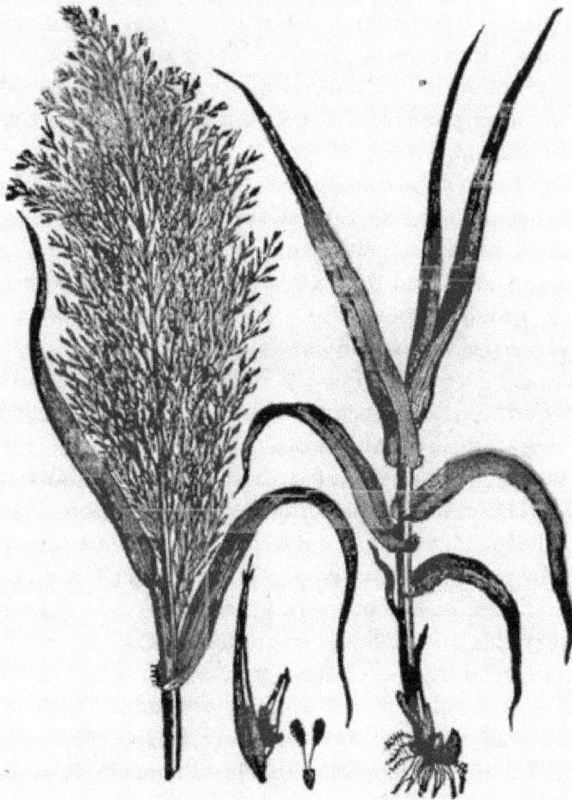

REED.

: ROSE. A genus of plants, the stems of which are generall, prickly, and the leaves pinnate. Of the rose there are hundreds of varieties, ranging in size from a tiny bush to a tree or immense vine. No flower has more varieties than the rose, and it is capable of still greater variation by cultivation. The varieties of shape, color and fragrance are almost endless. The "rose of Sharon" (Sol. Song 2. 1) is supposed to be the

Rosa Centifolia, or hundred-leaved rose, a variety much prized by the ancients. From time immemorial the rose has been an emblem of beauty and intrinsic worth (Isa. 35. 1).

RYE. The word *rye* occurs twice in the Scriptures, and it is considered doubtful whether the same grain is referred to in both places. It is supposed that the "rye" of Isa. 28. 25 is the same as that now known by that name (*Secale Cereale*). The "rye" of Ex. 9. 32, however, is supposed to be the *Triticum Spelta,* a species of bearded wheat, much cultivated in Egypt in ancient times.

RUE (*Ruta Graveolens*) occurs in Luke 11. 42. This is a perennial plant, growing, from two to three feet high. The leaves are divided, or doubly pinnate. The bark near the base is rough and woody, and the stem terminates in several branches, ending in smooth, green twigs. It bears a small, yellow flower. The plant is used as a stimulant and condiment.

SAFFRON (*Crocus Sativus*). This plant is mentioned but once in Scripture. It occurs in Sol. Song 4. 14, in connection with other fragrant substances. The saffron is a perennial plant, having a somewhat flat, bulbous root. The stem is a long and tender tube, terminating with flowers, which are generally yellow. A very fine color is obtained from this variety. The whole plant is also medicinal. Although the plants now found in Palestine are not very odorous, yet it is asserted that it is susceptible of a delicate fragrance by cultivation.

SHITTIM=SHITTAH. These words, which are synonymous, are left untranslated in our English version. The Ark of the covenant was made of this wood (Ex. 25. 10); the staves (Ex. 27. 7); the table (Ex. 30. 10); tabernacle (Ex. 26. 15). (*Shittah :* see Isa. 41. 19). In the different versions, where the words are translated, the renderings are not at all uniform. The German version renders it fir. The best and most recent authorities concur in rendering it *acacia*—probably *Acacia Seyal.* This tree, it is known, was obtainable at several points of Israel's wilderness route. Its wood is both durable and beautiful.

SPIKENARD (Sol. Song 1. 12; Mark 14. 3; John 12. 3). A very fragrant and aromatic plant, known to botany as *Nard-*

ostachys Jatamansi. It was used as a perfume, and also for embalming purposes. It is the same as *Nard,* already described, which see.

SWEET CANE (Isa. 43. 24; Jer. 6. 20). From its connections we learn that this substance was brought from "a far country," and is supposed to be identical with the fragrant *Calamus,* already described.

SYCAMINE (Greek, *Sukaminos*) occurs only once (Luke 17. 6): "If ye had faith as a grain of mustard seed, ye might say unto this sycamine tree," &c. This is merely the untranslated Greek name for the mulberry tree (*Morus Nigra*), already described.

SYCAMORE (*Ficus Sycamorus*). The sycamore is a species of fig-tree, differing from the true fig in some important particulars. The sycamore is a very large tree, with very extended branches, and is, on this account, planted as a shade-tree. Its leaves resemble those of the mulberry; hence its name from *suka,* a fig, and *moros,* the mulberry-tree (literally, fig-mul. berry). The fruit is sweet and good, but the wood is light and porous, and of little value.

TARE. There is necessarily some uncertainty as to what particular plant our Saviour refers to in the parable of the sower (Matt. 13. 25, &c.) There are several plants native to Palestine that answer the description very well. There are several varieties of *ervum,* very troublesome to wheat-growers. One kind, the *ervum hirsutum,* or hairy ervum, has been known to overgrow and destroy entire fields of wheat. Its seed is contained in small pods. The plant is tender, and eaten by cattle. The weight of opinion, however, inclines to the *darnel* (*Lolium Temulentum*) as the most probable plant. It resembles the wheat somewhat, and is a great detriment to its growth.

TEIL-TREE (Isa. 6. 13) (*Tilia Europæus*). Commonly known as the linden-tree in Europe and America. The *tilia* is a very tall and graceful tree. Its leaves are deciduous, heart-shaped and serrated. The flowers of the *tilia* are odoriferous and contain much honey, which is eagerly sought by bees. Although the wood is very soft, it is much valued for manufacturing purposes. The teil is represented in America by the basswood-tree, which is the *Tilia Americana* (American teil).

THORNS, THISTLES. Hebrew writers assert that there are twenty-two different words in the Scriptures indicating thorny and prickly plants; hence the difficulty of identifying them all will be apparent, and especially so when it is remembered that they are all rendered by the general terms *thorn* and *thistle.* It may, however, be of interest to the reader to know that the word *akantha*, rendered *thorn* in Matt. 7. 16; 13. 7; 27. 29; John 19. 2, 5, is supposed to refer to the *Zizyphus*, a small tree having leaves resembling the ivy, and of a deep, glossy, green color. Its branches are soft and pliable, and covered with small, sharp spines. Moreover; being abundant, it is supposed that the "crown of thorns" which was placed on our Saviour's head was made of the pliant branches of this tree. Botanists have, therefore, named it *Zyziphus Spina Christi* (thorn of Christ).

THYINE WOOD. Occurs only in Rev. 18. 12, where it is spoken of as one of the valuable things which should cease to purchased upon the downfall of Babylon. Thyine wood is believed to be the *Thuja Articulata*, a tree found in the neighborhood of Mount Atlas. Its wood is balsamic and odoriferous, and was anciently much employed for ornamental purposes.

WILLOW (Job. 40. 22; Psa. 137. 2; Isa. 44. 4). A genus of trees and shrubs now found in all warm and temperate regions of the globe. There are scores of widely-different species, and a description of them cannot be given here. The willow upon which the Jews "hung their harps by the rivers of Babylon" (Psa. 137. 2) is supposed to be the common "weeping willow" (*Salix Babylonica*) (willow of Babylon). It was introduced into Europe, from the East, centuries ago, and thence into America. Its leaves are lanceolate, and branches pendant. When growing in a warm, moist soil, it attains a gigantic size.

WHEAT (*Triticum*) is of frequent occurence in the Bible. It is the most valuable cereal grass known to man, and has been cultivated from remote antiquity. Since the cultivated varieties in the East do not differ materially from that known to all our readers, a description seems unnecessary. A certain species, of which the reader may not be familiar, is that called

Mummy wheat (*Triticum Compositum*); so named because
reproduced from seeds found in an Egyptian mummy-case supposed to be three thousand years old.   The ears, or heads, are
not single, but have shoots, or spikelets, sometimes eight or ten
in number.   Although prolific, for various reasons it is not
considered as valuable as our ordinary varieties.

WALNUT.   Although this word does not occur in the Scriptures, it is supposed that the word *Egoz*, rendered "nuts" in
Sol. Song 6. 2, refers to this tree.   The walnut (*Juglans*), in its
several varieties, is plentiful in the East, and is noted for the
excellency of its nuts and wood.

WHEAT.

WORMWOOD (Deut. 29. 18: Prov. 5. 4; Jer. 9. 15; Lam.
3. 15; Amos 5. 7) is the *Artemisia Judaica*, a perennial plant
found in Palestine.   It is more bitter than the European
species (*Artemisia Absinthium*).   The leaves are small, and
of an ashen color.   Its flower is small and yellow.   The whole
plant is medicinal, and is one of the most important plants
known to medicine.

QUADRUPEDS.

APE. "Once in three years came the navy of *Tharshish,* bringing gold and silver, ivory and *apes* and peacocks" (1 Kings 10. 22, also 2 Chron. 9. 21). It is quite evident from the above, that the *Ape,* was imported from a distant land. Authorities differ very much as to the location of *Tharshish.* Several writers hold that it is identical with Spain in Europe, while the weight of authority seems to be in favor of India. Among the strongest arguments are the following: 1. All the above imports were abundant there. 2. India is the original home of the *Peacock.* The Ape is tailless, larger than the monkey and generally smaller than the baboon. There are many widely differing species. For what purpose they were imported is not known. The worship of apes and monkeys was very common among pagan nations, and is still practiced to some extent. It is said that an ape's tooth was kept in a temple in Ceylon, and was regarded with extraordinary veneration, and immense wealth was accumulated by the continued offerings of the worshipers, but the temple was plundered and the tooth carried away by the Portuguese in A. D. 1554.

ASS, (*Equinas Asinus*). A quadruped generally classed by zöologists in the horse family, but it differs from the horse in many particulars. It is smaller in size, and differs in the shape of its body. Its head is large; with an arched facial

line. Ears long, mane erect. The hairs of the tail are short
and terminate in a tuft. There is also a streak of hair differing from the rest of the body in color,—generally grey, running on the back from the neck to the tail. This stripe is
crossed by another of the same color on the shoulders. Its
voice is a harsh, distressing bray (Job 6. 5; 30. 7). The ass was
used for agricultural purposes (Ex. 23. 12; Deut. 22. 10; Isa.
22. 20). For travel Gen. 22. 3; Num. 22. 23; Matt. 21. 2. It
was predicted that Christ should make his triumphal entry into

ASS.

Jerusalem on an ass (Zech. 9. 9), fulfilled (Matt. 21. 5). 1000
she asses owned by Job (Job 42. 12). The reference in Judges
5. 10, to a breed of white asses is corroborated by travelers.
"From early ages a white breed of this race was reared at
Zobier ❋ ❋ ❋ from which place civil dignitaries still obtain
their white asses and white mules" (*Kitto.*) The ass is not
improved much by domestication like the horse. In its wild
state it is larger, more graceful, and more fleet of foot. A very
fine species of a silver grey color is found in Africa. Reference is made to wild asses in Job 24. 5; Psa. 104. 2; Isa. 32.
14; Dan. 5. 2.

BADGER (*Meles Taxus*). "Thou shalt make a covering for
the tent of rams' skins dyed red, and a covering above of

Badgers' skins" (Ex. 24. 14, also Ex. 35. 7; 36. 19; 39. 14; Num. 4. 6 and Ez. 16. 10).

The badger (*Meles Taxus*) is an omniverous animal of the *Linnœan* genus, *Ursus*, or bear family. The badger is about two and a half feet in length, and a foot high when full grown. Its color is greyish brown above, and black beneath. Its head is white, with a black stripe lengthwise on each side. The hair is coarse and reaches to the ground, when the animal stands erect. The American badger, so numerous in the far West, is known in zöology as *Meles Labradorica.* It is smaller and

BADGER.

very different in its habits from *Meles Taxus.* Some eminent authorities (among them Kitto) hold that the word *tacash* does not denote badger as our Version has it, but the skin of some maritime animal, holding that it is not a native of Western Asia. This we think is an error. Luther and other translators uniformly render it badger (German "*Dachsfellen.*")

BAT (*Vespertilio*) (Lev. 11. 19; Deut. 14. 18; Isa. 2. 20). An animal of the class *Mammalia*, called by the Greeks *Cheiroptera*, (Hand winged). The bat suckles its young like other animals, but differs from them in having the power of flight. Upwards of 130 species of bats have been described by naturalists. Those of Europe and America are of the smaller species. The Bat is found in all parts of the globe, save in very cold regions. It is nocturnal in its habits, and frequents damp, dark places. In the mosaic code it was placed in the list of unclean animals.

BEAR (*Ursus*). A widely distributed genus of mammalian quadrupeds, embracing many species. The genus *ursus*, not only embraces the true bear, but also the glutton, wolverine badger, raccoon, &c. The true bear, however, is an animal of the Order, *Feræ*, Sub. Order, *Carnivora*, and tribe *Plantigrade*. Probably the largest species is the *Grizzly* (*Ursus ferox*) of North America. It is the most powerful and dangerous animal on the Continent. Its body sometimes attains a length of nine feet. The white or polar bear (*Ursus Maratimus*) is almost as large as the grizzly, but is less fierce, and has long, smooth, white fur, long neck and flat head.

The common black bear is *Ursus Americanus*. It is seldom more than five feet in length, and seldom attacks man. The yellow and cinnamon bear are only varieties of this species.

The bear of Scriptures is doubtless *Ursus Syricus*. It differs from other species in having a stiff mane, and erect hairs between the shoulders. Its color is light brown. Reference to the bear's affection for its young is made in 2 Sam. 17. 8; Prov. 17. 12; Hos. 13, 7, 8, bears attacked flocks (1 Sam. 17. 34, 35). Forty-two bad children of Bethel were slain by two she-bears for mocking Elisha (2 King 2. 24).

BEHEMOTH. For a description of this remarkable animal (See Job 40. 15–24). It is exceedingly difficult to determine what animal is meant in this sublime description, and consequently nearly all European translations of the Bible do not venture a rendering of the Hebrew word "*Behemoth.*" Several excellent authorities, among them Bochart and Calmet, suppose the *Hippopotamus*, or river horse, is intended, while others again contend that it is the *Elephant*, while the graphic description of Job applies in part to either animal, it fails when applied to any of them separately. Hence some maintain that it refers to some large quadruped now extinct. We will give a brief description of the *Hippopotamus*, so that the reader may judge for himself as to its identity with Job's "Behemoth."

There are two species of this animal, *Hippopotamus Amphibius* and *H. Liberiensis*. The latter is a much smaller species than the former, and naturalists have but recently become acquainted with it. The *H. Amphibius* is but little

BEHEMOTH.

inferior in size to the elephant. Its very short legs, however, give it a much smaller appearance. It is an aquatic animal, and never ventures far away from the water. The head is large, ears small, eyes small, and placed high so that the animal can see quite well without raising much of his body above the water. The feet have four toes each. Its front teeth are very large, and so set that he can bite off grass as though it were cut with a scythe. It has the thickest skin of any

animal known, being more than two inches on the back and
sides. Its tail is small, and out of proportion to the body.
The animal is of a brown color, and destitute of hair. It
secretes a thick oily fluid from its skin. The animal breathes
very slowly and hence it is able to remain under the water a
long time. For so large and unwieldy an animal it swims and
dives with great ease. Being a vegetable feeder it sometimes
does great damage to growing crops in the vicinity of water-
courses.

The elephant will be described in its appropriate place.

BOAR (wild) *Sus Scrofa.* Only once distinctly alluded to in
the Bible, "The boar out of the wood doth waste it" (Psa.
80. 13).

This quadruped is supposed to be the progenitor of the
domestic swine, from which it now differs considerably. The
color of the wild boar is a greyish black. Its hair is short and
wolly, with stiff bristles intermingled. They are much longer
and thicker on the top of the neck, forming an erect mane.

Its tusks are very long and formidable. This animal was
formerly abundant in Europe, but is now seldom met with
except in the extensive royal forests and preserves on the continent. It is still abundant in some parts of Asia, especially in
India. The boar is not considered a dangerous animal if
unmolested, but if pursued and attacked he is very fierce and
formidable.

The wild boar does great damage to growing crops, sometimes whole fields are destroyed in a single night by a drove of
these animals. Hence the Psalmist's allusion as above.

CAMEL (*Camelus*). Very frequently mentioned in the Bible,
and the most important and useful animal in the East. The
camel belongs to the *Tylopoda* (cushion footed animals), and
are classed in the order *Ruminantia.* There are only two
species of the *Camelus,* proper. The Arabian, (*Camelus
Dromedarius*) or one humped camel, found chiefly in Arabia
and adjacent countries, and *Camelius Bactrianus,* or Bactrian
Camel, having two humps. There is perhaps no other animal
in the world that more clearly demonstrates the wisdom and
benevolence of God toward mankind than this. The camel

seems to have been created with a special view to the country
it inhabits. Its whole structure seems to have been planned
with reference to the climate, soil, and food supply of this part
of the world. Its feet are admirably adapted to traverse the
soft, hot sands of the desert, consisting as they do of two toes,
soft and cushion-like, and terminated with a small nail-like
hoof. Its legs and neck are very long. Its power to resist
long-continued drought is wonderful. A portion of the paunch,
and the lining of the second stomach is honey-combed with
cells, in which the animal preserves vast quantities of water for
future use. It is said by some to be able to subsist a whole
month without drinking.

They can abstain from food almost equally well. All its
organs pertaining to food are constructed to meet privations
and emergencies. The coarse and prickly herbage of the
desert is eaten by them.

The most wonderful provision for long abstinence from food
is the hump on the camel's back. This hump is not caused, as
is supposed by some, by the curvature of the spine (its back-
bone is almost as straight as in other animals); but it is simply
a great hump of fat. The hump is a reserve of vitality, and it
is in a great measure absorbed in the system during a long-
continued journey. Acting on a knowledge of this fact the
Arabs are said to judge of the animal's capabilities by the con-
dition of his hump. The camel's nostrils are so constructe:
that it can close them at will, thus being able in a measure to
protect itself from the gusts of hot air laden with sand.

As this animal bows or kneels down to receive its burden
there are hard callosities on its breast and joints. Without this
provision of nature the flesh of the animal would be bruised in
raising or lowering its immense load.

Of the two species the Bactrian is by far the stronger. It
easily carries a burden of 1000 pounds, but must sometimes
carry much more.

Camels are first mentioned in Gen. 12. 16, Jacob receiving
some from the hands of Pharaoh as a gift. Job possessed
3000 camels before his great misfortunes, and 6000 after
his restoration. Rebekah rode on one on her way to meet her

future husband Isaac (Gen. 24. 61–64). The Ishmaelites used camels on their mercantile trips to Egypt (Gen. 36. 25). Although the flesh of the camel was—and still is—eaten by surrounding nations, it was unclean to the Jews (Lev. 11. 4). The camel is so frequently referred to in the Bible that a reference cannot be made to all the passages. The Hebrew word *Kirkârôth* in Isa. 66. 20, translated 'swift beasts," is supposed to refer to the *Dromedary*, a finer and lighter breed or variety of the Arabian camel. It is said that this variety can make a journey of more than a hundred miles in a day.

CAT (*Felis*). It is somewhat remarkable that this familiar animal is nowhere alluded to in the inspired books of the Bible. It occurs however in the *Apocrypha* (Baruch 6. 22), where it is mentioned as one of the animals that sat on the Idols of Babylon. Although the domestic cat is now quite common in Palestine, it is probable that it was very rare, or perhaps unknown to the Jews in ancient times.

CONY (*Coney*). Hebrew *Shôphâns*. The best authorities agree that this animal is the *Hyrax* of Zöology. There are several species of this animal, and that mentioned in Scripture is doubtless *Hyrax Syracus* (Syrian Hyrax). The *Hyrax* is a very odd and interesting animal, although not much larger than the rabbit; it is structurally almost identical with the huge *Rhinoceros*. The great naturalist *Cuvier* says of them, "Excepting the horns they are little else than rhinoceroses in miniature." This singular animal is about a foot long, and about ten inches high. Its hair, which is of a brownish color, is interspersed with bristles. It has four toes on each of the forefeet, and but three on each of the hind feet. It has a mere trace of a tail. Although it lives among the rocks (Psa. 104. 18), it does not burrow, its feet not being adapted for it. They are very timid and perfectly harmless. Solomon calls them a "feeble folk" (Proverbs 30. 26). They were classed by Moses among the unclean animals (Lev. 11. 5; Deut. 14. 7).

CHAMOIS (Deut. 14. 5). The Hebrew word *Zemer* is variously translated in the different versions. The *Septuagint* and *Vulgate* have it *Camelopard* (Giraffe), Luther has it *Elend* (the moos or Elk), while the English has it *Chamois.* The word must, therefore, be incorrectly rendered in several of the above versions. The best authorities agree that the *Zemer* of Deut. 14. 5, is the *Ovis tragelaphus* of zöology. The *Camelopard* is strictly a South African animal. The *Moos* or *Elk* does not venture so far south, while the true *Chamois* is not native to Palestine. The *Ovis tragelaphus,* however, is found in the vicinity of Palestine. It is known as the "Syrian sheepgoat" (J. F. & B.) It is an animal occupying an intermediate place between the sheep and goat (Kitto). Its neck and whole front to the knees is covered with long yellowish hair.

Dog (*Canis*). There seems to have been different varieties of the dog in ancient, as well as in modern times. The animal was well known to the Jews both in its wild and domesticated state. The dog is not so highly esteemed by the Orientals as by Europeans, and nearly all the references to it in the Bible are of an aversive character. There seems, however, to have been some value set upon it (Deut. 23. 18).

They were also trained to watch flocks (Job 30. 1), accompanied their masters on a journey (Tobit 5. 16; 11. 4); wild dogs of various breeds are still met with in Eastern countries, and even in large cities there are packs of dogs that have no owners. They prowl about the streets and make night hideous by their howling as in ancient times (Psa. 59. 6-14). These wild dogs still follow caravans as in Ex. 22. 31.

Dromedary. The Arabian Camel (*Camelus Dromedarius*), is sometimes, though erroneously, called the Dromedary. Strictly speaking the Dromedary is an improved breed of the Arabian or one-humped camel. A certain author says, "It bears the same relation to the camel that the fleet racer does to the cart horse." Its limbs are more slender than the ordinary camel, and it is characterized by greater symmetry and beauty. Its powers of endurance are extraordinary. It will travel with a rider on its back for twenty-four hours without stopping, at a very rapid rate. It can travel 600 miles in five days (Cham. Enc). (See Jer. 2. 3; Isa. 60. 6; Esth. 8. 10; 1 Kings 4. 28).

Elephant (*Elephas*). This animal is not mentioned in the inspired portion of the Bible, but is found in 1 Macc. 6. 34, &c.,

where it is said that there "were strong towers of wood, which covered every one of them, and were girt fast under them with devices. There were also on every one of them two and thirty strong men that fought upon them besides the Indian that ruled them."

The employment of the elephant in this manner for military purposes was, and still is, common in Oriental lands. Although the elephant was not a native of Palestine, and probably little known to the Jews, a brief description seems to be in place.

The elephant is the largest terrestrial quadruped known. The ordinary height is eight feet, but often exceeds ten feet, and it often exceeds five tons in weight.

DROMEDARY.

The elephant's head is large, neck short, ears large and pendant. Its legs are almost straight and very thick. Its skin is very thick, and dark brown in color, and almost destitute of hair.

Its sense of smell and hearing is very acute. It takes all its nourishment, both food and water with its trunk. The water is first drawn into the trunk and then discharged into the mouth. The trunk which is about eight feet in length, is one of the most wonderful things known to man. Cuvier, the great naturalist, says there are over 40,000 muscles connected with it. Although it can coil it about a tree and wage fearful destruction with it, yet it is so constructed that it can pick up minute objects with it, untie knots, uncork bottles, &c.

The elephant is the most intelligent animal known. Some of its actions seem to be, not the result of instinct or intuition, but of abstract reasoning.

There are two species of this animal: *Elephas Indicus* (Indian Elephant), and *Elephas Africanus* (African Elephant).

FALLOW DEER. All authors consulted agree that the Hebrew word *"Jachmûr"* does not denote the fallow deer. The German version renders it *Büffel*, (buffalo or bison). This seems to be equally incorrect. Naturalists are generally agreed that the *Jachmûr* of the Bible is the *Oryx leucoryx*, now known to science as *Antelope leucoryx*. This animal is still native to

FALLOW DEER.

the regions adjacent to Palestine, and is called by the Arabs *"Jazimûr,"* which is identical with the Hebrew *"Jachmûr."* This animal is a species of antelope. Its color is white, with black cheeks, and a black patch on the nose. In size, it is larger than the goat. Its horns are long, annulated and curved slightly backward. It was a clean animal (Deut. 14. 5), and its flesh royal food (1 Kings 4. 23).

4

FERRET (Lev. 11. 30). Much uncertainty exists, as to the correctness of the English rendering of the word *Anakah.* There is no uniformity in the translation of this word in the various versions. The German has it "*Igel*," (hedgehog), an animal differing widely from the *ferret.* Eminent linguists and Bible zöologist maintain that the word indicates a species of newt or lizard. Prof. C. H. Smith, an eminent authority on the natural history of Palestine, says, it is a lizard known to naturalists as *Lacerta gecko.*

The ferret (*Mustela furo*) belongs to the weasel family. It is more than a foot in length. Eyes of a pink color. The color of the fur is yellowish, with more or less white. Its form is slender. It is capable of domestication and very useful in destroying rats and other vermin.

FOX (*Vulpes*). The fox is frequently mentioned in the Bible, and it is agreed by all authorities that this term does not always denote the true fox but also the *Jackal,* which will be noticed in its proper place. The fox belongs to the same genus as the dog (*Canidæ*). It differs from the true dog very essentially. It is lower in stature in proportion to its length than other animals of this genus. The head is round with a pointed muzzel. Its ears are short and triangular. Its tail is heavy and bushy. Limbs slender. They burrow in the ground or secrete themselves in the crevices of rocks. They are remarkable for their speed, cunning and acute sense of sight, and smell. The different species of this animal differ widely in color, form, and habit.

Sampson caught 300 (Judges 15. 4). Prophets likened to foxes (Ezek. 13. 14). Applied derisively (Nehemiah 4. 3). Referred to by the Saviour (Matt. 8. 20; Luke 9. 58). Used figuratively (Luke 13. 32).

GOAT (*Capra*). A genus of mammalian quadrupeds found in temperate and sub-tropical countries. There are many varieties of this useful animal, differing very widely in their appearance. Of this animal there were probably four varieties known to the Jews. "1. The domestic Syrian long-eared sheep, with horns rather small and variously bent. The ears

very long an pendulous. Hair long, and sometimes black.
2. The *Angora*, or *Anadoli*, breed of Asia minor, having very
long and fine hair. 3. The Egyptian breed, with long spiral
horns. Long brown hair, and long ears. 4. A breed from
Upper Egypt, without horns, having the nasal bones singularly
elevated. The nose contracted with lower jaw protruding the
incisor teeth," (Kitto). The goat was a pure animal to the
Jews, and used for sacrifice (Ex. 12. 5). Its skin was manufactured into a bag, called a "bottle," for containing liquids
(Matt. 9. 17). Its hair was made into cloth (Ex. 35. 23, &c.)

GOAT.

Its skin made into clothing (Heb. 11. 37). The wicked compared to goats (Matt. 25. 32). Wild goats, (1 Sam. 26. 2; Job.
39. 1; Psa. 104. 18).

GREY-HOUND (Prov. 30. 31). A species of dog noted for its
gracefulness of form and extraordinary fleetness. Its power
of scent, however, is not so acute as in the ordinary dogs, and
it pursues its game principally by sight.

HARE (*Lepus*). A *Linnæan* genus of rodent quadrupeds, of
which there are many species, inhabiting almost every quarter
of the globe. There are two species of hare with which the
Jews were probably well acquainted. *Lepus Syriacus*, (Syrian
hare) and *Lepus Sinaiticus*, (hare of Sinai). They differ
slightly in form and color. The hare differs from the rabbit, in
being larger and its hind legs being longer in proportion to its
body. It is strictly herbiverous.

The hare is not a ruminant (cud chewing) animal as might be inferred from Lev. 11. 6, and Deut. 14. 7. All true ruminants have four stomachs, and the feet bisulcate (cloven-footed)

GREY-HOUND.

while the hare has but one stomach, and does not raise and re-chew its food. It has been observed, however, by naturalists that it is in the habit of reserving food in the hollow of its cheek, which it maunches at its leisure. Their jaws are, therefore, more or less in motion from this cause, and also in grind-

HARE.

ing their incissor teeth upon each other, in common with other rodents to prevent their growing too long. It was probably on this account that it was classed with the unclean animals.

HART. A species of deer supposed to be the *Cervus Bar barus*, or Barbary Deer. This animal has a general resem-

blance to the European fallow deer, but differs from it by the
want of absantler, or second branch on the horns counting
from below; and in its spotted appearance which disappears,

HART.

when three, or four years old.
Its flesh was used on Solomon's
table (1 Kings 4. 23). Its well-
known propensity for water
spiritualized (Psa. 42. 1). Its
fleetness (2 Sam. 22. 34; Psa.
18. 33). The gentleness and
affection of the female (Prov.
5. 19). The well-known habit
of the deer in concealing its
fawn for some time after birth is alluded to in Job 39. 1. It
was classed with the clean animals (Deut. 12. 22; 14. 4, 5).

HIND. Of the hind it is only necessary to say that it is the
female of the hart (2 Sam. 22. 34; Hab. 3. 19; Job 39. 1;
Prov. 5. 19; Cant. 2. 7; Jer. 14. 5).

HEDGEHOG. Although this animal is not mentioned in the
English version, it is held by some eminent scholars that
the Hebrew word "*Kippo*" in Isa. 14. 23; Isa. 34. 11, and
Zeph. 2. 14, rendered "Bittern" in the English version, really
denotes the hedgehog, or *Porcupine*. The argument resting
on the first syllable of "*Kippo*," which denotes *spine*. Although
Luther always renders it hedgehog (*Igel*), other versions are not
uniform. The Arabic has it "*Bustard*," some others have it
"*otter*" and "*Osprey*." Although nearly all commentators
give "hedgehog" as the preferable rendering we believe it to be
inadmissible. Even though "Kippo" etymologically considered might call for such a rendering, yet when the word is
considered in its relation to the passages in which it occurs it
is quite evident that "hedgehog" will not answer.

The hedgehog cannot climb. It does not frequent cold damp
ruins, it is not aquatic. It prefers dry places, we believe the
English translation to be correct. (See Bittern).

HORSE (*Equus*). A genus of quadrupeds now found in
almost every inhabitable region of the globe, and by far the
most useful animal known to man for general purposes. It

HEDGEHOG.

seems unnecessary to give a description of this well-known
animal. It is supposed to be a native of Central Asia, and
from thence to have been introduced into Egypt, where it was
first domesticated. It is found sculptured on Egyptian ruins,
at least 4000 years old, generally in hunting and battle scenes.
The horse is first mentioned in Gen. 47. 17, where Joseph
receives them from the Egyptians in exchange for bread. The
Egyptians accompanied the Jews at the funeral of Jacob in
Canaan with horses (Gen. 50. 9).

In the early history of the Jews, the horse does not seem to
have been highly esteemed, as the law of Moses (Deut. 17. 16)
expressly forbade any king to " multiply horses." David ham-
strung the horses captured from the enemy in battle (1 Sam.

8. 4), but Solomon who seems to have been fond of horses, had 40,000 stalls of them (1 Kings 4. 26). It seems he imported a favorite breed from Egypt (1 Chron. 1. 14–17). In the New Testament there are but few references to the horses. Our Saviour does not once mention it.

HYENA. Does not occur in the inspired portion of the Bible, but in Ecclesiasticus 13. 18, (Apocrypha). "What agreement is there between the hyena and a dog." The hyena is a carnivorous quadruped included by *Linnæus* in the genus *Canis*, or dog family. It is now generally placed by naturalists in the genus *Viveridæ*. The hyena is larger than the dog, which it somewhat resembles. There are various species of different

HYENA.

colors. It is generally striped or spotted with a ridge of stiff, coarse, erect hair on the back and neck. It is exceedingly foul and loathsome. Subsisting almost entirely on carrion, and even robbing graves for the dead. The species found in Palestine is the striped hyena (*Hyena Vulgaris*). In 1 Sam. 13. 18, where occurs "Valley of Zeboim" the original is *Ge-hat-tseboim* (valley of hyena).

JACKAL. Although this word does not occur in the authorized version, it is nevertheless certain that it is referred to in several passages. Linguists are agreed that the word *Shual*, (plural *Shualim*), uniformly rendered foxes, is not specific, but generic in its character, and not only means foxes

proper but also animals allied to them.   The Jackal is a species
of wild dog, quite common in the East.   It is generally of a buff
color, about the size of an ordinary dog.   Its head is narrow,
muzzle-pointed, tail bushy, tipped with black hair.   They are
very foul in their habits, prowling about at night in quest of
food, howling in a hideous manner, most commentators agree
that the 300 "foxes" that Sampson caught (Judges 15. 4, 5),
were jackals as this animal roams about in packs, large num-

JACKAL.

bers could be more easily captured while the fox is a solitary
animal.   In Psa. 60. 10, where David in speaking of his
enemies, says, "they shall be a portion for foxes." *Shualim*
clearly denotes *Jackals*, as they are very prone to rob graves
of their dead, a thing which foxes never do.

LEOPARD (*Felis Leopardis*). A carnivorous quadruped of
the cat tribe, and one of the largest animals of the genus *felidæ*.

Naturalists say it is almost identical with the *Panther*, being only variations of the same species.

The leopard is a large and dangerous beast, still found in the mountains of Lebanon and other regions of Asia and Africa. It is generally of a yellowish color, dotted over with spots of a darker hue, the spots being ranged in rows. Its form is very graceful, and like other animals of the cat tribe is noted for its agility and cunning. The leopard used as an emblem of conquest (Dan. 7. 6). Its cunning and watchfulness (Jer. 5. 6 and Hosea 13. 7). Its-spots referred to (Jer. 13. 23). (See also Cant. 4. 8; Isa. 11. 6; Hab. 1. 8; Rev. 13. 2).

LEVIATHAN. The meaning of this untranslated Hebrew word is quite obscure. It occurs five times in the Sacred text, and is left untranslated in every case in the English version except in Job 3. 8, where the word is rendered mourning, and the original leviathan is put in the margin. Luther gives the passage literally, "*Und die da bereit sind zu erwecken den Leviathan*," (And they who are ready to raise up the Leviathan). In Psa. 74. 14, and 104. 26, Luther makes it "*wallfish*," (whale). It also occurs in Job 41. 1, and Isa. 27. 1.

It is quite probable that Leviathan is a general term, denoting any great monster. In Job 41, where an extended description is given of it, the Nilotic Alligator is probably referred to. In the Psalm the whale seems to answer the description best.

LION (*Felis Leo*). Rightly named "the king of beasts." This most formidable of all animals, is a carnivorous quadruped of the cat tribe (*felidæ*). Its color generally is a tawny yellow. Its usual length from the nose to the tail is eight feet. But the African lion often exceeds that length. Its tail is about four feet in length, terminated by a tuft. He has a heavy, shaggy mane, his eye is quick and flashing, and like all feline animals he sees well at night. The lion is very muscular and strong, and it is said that he can carry off a heifer as easily as a cat does a rat. He is not so large an animal as one, on account of his strength, is led to suppose, as the Asiatic lion weighs about 450 pounds, while the African species is somewhat heavier.

According to ancient historians the lion was once an inhabitant of Eastern Europe, but he has been exterminated for many
centuries, and is no longer met with in Syria and Palestine,
where he was once abundant. There are many references to
the lion in Sacred history. He was made an instrument of
the wrath of God in the destruction of multitudes of people
(2 Kings 17. 25, 26). A great dread to shepherds on account
of his depredations (Isa. 31. 4). In the early period of the
Jewish nation, lions had their dens in the underbrush on the
banks of Jordan, and became unusually fierce when driven
·from their haunts by the annual overflow (see Jer. 49. 19; Jer.
4. 7). His ferocity (Nah. 2. 12). Reference made to a mode of
capturing him alive (Ezek. 19. 8, 9). An emblem of cruelty (Psa.
7. 2; Psa. 22. 21; 2 Tim. 4. 7). A type of Satan (1 Peter
5. 8). Daniel's adventure in the lion's den (Dan. 6. 16.)
Sampson slays one (Judges 14. 5); also David (1 Sam. 17. 34).

MOLE (*Talpa Europœa*). A small quadruped of the order
Insectivora. It is a burrowing animal, and lives almost
entirely under ground. The ordinary variety is about the size
of the mouse. Its eyes are very small. Its feet are broad,
and adapted for burrowing. Its hair, which is generally black,
is short and silky, and stands perpendicular to the skin, thus
enabling the animal to move backward, or forward with equal
ease. It burrows very systematically and its underground
galleries are interesting subjects for study. The shrew mole
(*Scalaps*) and star nosed mole (*Condyluria cristata*) are
American species.

The mole was proscribed by the mosaic law as unclean (Lev.
11. 30), used figuratively (Isa. 2. 20).

MOUSE. A small quadruped of the genus *mus*, which also
includes rats. It is presumed that this animal is so familiar to
the reader that a description seems unnecessary. There are
many species of this animal, the largest (*Mus Barbarus*) being
almost the size of the common rat, while the smallest (*Mus
Messorius*) is scarcely more than two inches in length. The
Mus lencopus, or white-footed mouse is found only in America.

The mouse was an unclean animal to the Jews (Lev. 11. 29;
Isa. 66. 17).

MULE. Mentioned quite often in Scripture. It is a *hybrid* animal, and a description is not necessary. The mule was produced already in the earliest ages of mankind, as they are figured on the oldest Egyptian monuments. They were anciently much more highly prized than now (See Gen. 36. 24; 1 Kings 10. 25; 2 Kings 5. 17; 1 Chron. 12 40; Ezek. 2. 66; Neh. 7. 68; Esth. 8. 10; Isa. 66. 20).

Ox. Several species of bovine quadrupeds were reared by the Jews. One species of ox more nearly resembled the *bison*, having long downward curved horns, with a considerable elevation of the body at the shoulders forming a sort of hump.

Assuming that the oxen used by the Jews did not essentially differ from our domestic species, a description of them seems unnecessary. The ox was held in high esteem by all Eastern nations. It was employed more than any other animal in agriculture by the Jews.

They were used for ploughing (Deut. 22. 10).

" Elisha, the son of Shaphat was ploughing with twelve yoke of oxen before him " (1 Kings 19. 19). Constituted part of Job's wealth (Job 1. 14, 15).

They were used for treading, or thrashing grain (Deut. 25. 4; Hos. 10. 11). Were used for drawing (Num. 7. 3). Sometimes used as beasts of burden (1 Chron. 12. 40).

Laws concerning vicious bulls (Ex. 21. 28, 29). The bull was an object of worship by the Egyptians, and hence when the Jews departed from the worship of God they reverted to that form of idolatry (Ex. 32. 4, 5; 1 Kings 12. 28).

PYGARG (Deut. 14. 5). This animal is a species of antelope, known as the *Oryx Addax*. It is about three and a half feet high at the shoulders. Its horns are wreathed, and about two feet in length. It differs from other species of antelopes principally in having a heavy tuft of coarse hair, (a sort of beard) under the gullet. The color of its hair is light grey on the head, neck, and shoulders, while the rest of the body is white. It is quite common in the regions traversed by the Israelites.

ROE. Known to the Hebrews as *Tsebi* and *Dorcas*, is a species of antelope commonly called the *Gazelle*. There are two species of this animal in the East, both probably known to the Jews. The *Ariel Gazelle* (Antelope Arabica), and *Antelope Dorcas.* They do not differ materially.

The Roe is an animal of great beauty. It is very timid and sensitive. It is seldom more than two feet high at the shoulders. Its eyes, which are black, are very large, soft and expressive. Its ears are long, narrow and pointed. Its horns are annulated, and by inclining first inward, and then outward, resemble the framework of a lyre, hence its horns are known as "*lyrated.*" It is generally of a tawny color with the under parts, white with a brown stripe on each flank. Its face which is of a redish color is also striped.

This animal is excelled by none other in gracefulness of form and movement. It is also very fleet (See 1 Chron. 12. 8; Prov. 6. 5, and 5. 9; Cant. 2. 7; 3. 5; 8. 14; Isa. 13. 14).

SHEEP (*Ovis*). A genus of ruminant animals found in most quarters of the globe. There are many species, sometimes differing considerably. It is presumed that the reader is familiar with the animal, so that a description is unnecessary. It has in all ages been one of the most useful animals to mankind, being perhaps the first animal domesticated (Gen. 4. 2). It was the first animal used for sacrificial purposes (Gen. 4. 4), and was emblematic of "the Lamb slain from the foundation of the world" (Rev. 13. 8).

The importance of this animal to the Jews may be seen in the fact that it is more frequently mentioned than any other in the Bible.

It doubtless constituted the chief wealth of the people in the earlier period of their history. Job had 14,000 of them (Job 42. 12) Under the Jewish ceremonial law it was the most important sacrificial offering, doubtless from the fact that its blood particularly was typical of the blood of Christ (See Isa. 53. 7; John 1. 29; Acts 8. 32; 1 Pet. 1. 19).

The term "Lamb" is applied to Christ twenty-two times in Revelations.

Christians compared to sheep (Psa. 95. 7; Matt. 25 33; John 10. 3, &c. ; 21. 16). Christ the Shepherd of the spiritual flock

SHEEP.

(Zech. 13. 7; Matt. 26. 13; Matt. 15. 24; John 10. 3; Heb. 13. 20). Read also Psalm 23.

SWIFT BEASTS (Isa. 66. 20). Reference is doubtless made to a variety of the *Camelus Dromedarius* or one-humped camel. It differs from the ordinary Dromedary in being lighter and

more symetrical in form. Of all animals the Dromedary
possesses the greatest powers of endurance. (See Dromedary.)
SWINE. Notwithstanding the prohibitory law (Lev. 11.
7), it is quite probable that some of the Jews partook of swine flesh
(Isa. 65. 4). It is spoken of as a familiar animal, and was
extensively used as food by nations adjacent to the Jews (Matt.
8. 30; Mark 5. 2; Luke 8. 33; Luke 15. 15). Metaphorically
mentioned (Matt. 7. 6).

This animal has already been described in the article
" Boar."

UNICORN. There is perhaps no animal mentioned in the
Bible, the identity of which is involved in greater obscurity
than the *Unicorn.* As its name implies, it is a one-horned
animal (*Lat. Unum Cornu*). But there is no such animal
known to zöology, and its existence is regarded as fabulous by
scientific men. The ancient Greek and Roman writers represented it as a native of India. It was larger and swifter than
a horse. The color of its body was said to be white, its head
red, its eyes blue. A long straight horn was said to project
from its forehead. Although Oriental travelers have heard of
one-horned animals, and although various one-horned animals
are figured on ancient monuments, their existence (with the
exception of the Rhinoceros) is considered legendary by eminent zöologists, Prof. C. H. Smith, an authority on oriental
zöology, says, "The radical meaning of the Hebrew word
furnishes no evidence that an animal, such as is now understood by *Unicorn,* was known to exist."

The rendering of the Hebrew word *Reem,* by *Monokeros*
(Greek, "one-horned,") in *Septuagint,* led the way for its
adoption in later versions. The *Vulgate* has it, *Rhinoceros.*
It is, perhaps, impossible to determine accurately what animal
is meant. Several authorities have supposed that a bovine
animal now known as *Bibos Cavifrons* is intended. Recurring
to the generally received opinion that a one-horned animal is
demanded, recourse must be had to the *Rhinoceros.* This is
the only " one-horned" animal of which we have any definite
knowledge, and some eminent authorities have accepted it as
the biblical unicorn. The *Rhinoceros* (Greek, "*nose-horned* ")

is a mammalian quadruped of the genus *Pachydermata ordinaria* of which there are eight species. It is a very large, heavy and clumsy animal, next to the elephant the largest terrestrial mammal. The head is very large, muzzle long and somewhat arched. Projecting from this arch is a single horn, which contrary to the horns of other animals springs from the skin. Several species have a second, but smaller horn above it on the forehead. Its legs are very thick. Each foot has three toes terminated by hoof-like nails. Its skin is very hard and thick, and in some parts of the surface, folded so as to permit the various motions of the body. The horn is a powerful weapon of offense and defense. For unicorn (see Num. 23. 22; 24. 8; Job 39. 9; Psa. 29. 6; Isa. 34. 7; Deut. 33. 17; Psa. 22. 21).

WEASEL (*Mustela*) (Lev. 11. 29). A small quadruped of the genus *mustela*, which embraces several widely differing species. The best-known species is the *mustela vulgaris* found in nearly all countries of the Northern hemisphere. The weasel proper is a very slender animal. Its legs are short. Its toes separate or parted. Its length is from seven to eight inches, its height hardly more than two and a half inches. Its eye is small, yet very sharp and black in color. The various species differ much in color in different quarters of the globe. Some authorities render the Hebrew word " *Choled*," "mole," instead of weasel in the above passage.

WHALE. "And God created great whales" (Gen. 1. 21). The largest creature known to man. A marine animal of the order *Cetacea*. Of this order there are several Genera, the *Balœnidœ*, or various species of *whalebone* whales. The *Physeteridœ*, or *sperm* whales. The *Delphinidœ*, such as Dolphins and Porpoises.

All animals of the above order are mammals. That is, they suckle their young (producing one at a birth). The reader will perceive that the term *whale* is very vague, indeed, and used in a collective sense to denote all the animals of this order. Since there are many widely differing species embraced in the term " whale," a description is out of the question.

·Critics have indulged in considerable discussion with regard to Bible references to the whale. It is held by some that the whale proper did not frequent waters navigated by nations of sacred history. More especially do infidels take advantage of the assertion of travelers and historians that it is not found in the Mediterranean sea. This they hold is proof that the narrative of Jonah and the whale is false. To this we may remark that the true whale has been seen in the waters of this sea even in modern times. Although it is very rare now, yet anciently it may have been abundant.

Again.—Infidels tell us that the whale was physically unable to swallow Jonah, its throat being very small. The occurrence being miraculous removes it above the question of physical impossibility In looking at the subject more closely, however, we see no grounds whatever for this alleged physical impossibility. In the original narrative (see Jonah, 1st Chap.) it is not said that a whale swallowed Jonah, but a "great fish," which God prepared (provided). Now, although the true whale cannot swallow a man, there are "great fish" in the sea now, that can do it. The *shark* can do it. Also a species of the *Acanthias*. True, our translators make Christ say that Jonah was swallowed by a whale (Matt. 12. 40), but a close examination of the original removes the difficulty at once. The original *Ketos* is used by ancient Greek writers to denote any sea monster, or huge fish (see Liddell and Scott's Standard Greek Dictionary). In the revised version, the translators have taken the precaution to qualify "whale" by adding "sea monster" in the margin.

WILD Ox (Deut. 14. 5). Most Hebrew scholars agree that the original indicates a kind of Antelope, of which there were several species in the regions known to the Jews. Excellent authorities agree that the particular species intended is the *Oryx tao* or *Nubian oryx*. It differs from the *Oryx Leucoryx* (already described) principally in being larger and more slender. Its horns also are more curved and longer. It is found in the desert, west of the Nile. Representations of it are found on ancient Egyptian monuments.

WOLF.

WOLF. An animal of the genus *Canis* found both in the old and new world. There are many species of this animal differing considerably both in nature and appearance.

The common wolf of Europe and Asia (*Canis lepus*) of a yellowish or tawny grey color. Its hair is coarse and thick. Its eyes obliquely set. Ears long and pointed. Its tail long and bushy. It roams in packs, is very fierce, and a source of great loss to settlements bordering on the forests, and mountains of Europe and Asia. Although now extinct, it was formerly abundant in Palestine (see Jer. 5. 6; Jno. 10. 12; Hab. 1. 8; Matt. 7. 15; 10. 16; Luke 10. 3; Acts 20. 29). The wolf must not be confounded with *wild dogs* and *Jackals*, which are still found in Palestine.

5

ICHTHYOLOGY.

FISH. From the many allusions to *fish* in the Bible, we infer that the Jews were well acquainted with the various kinds, with which Palestine, and adjacent countries abounded. Although the term fish and "fishing" occurs frequently in the Bible, yet there is nowhere any specific reference to a particular species, and hence our observations on the fish of the Bible must be of a general character. The River Nile abounds with fish of various kinds, and during its annual overflow, when its waters are distributed throughout Egypt by means of irrigating canals, its fish are brought within the reach of all. During their servitude in Egypt the Jews used fish as food, and during their wilderness journey, when their unbelieving souls "loathed this light bread," which God sent from Heaven (Num. 21. 5) they lusted after the flesh pots of Egypt and said, "we remember the *fish* which we did eat in Egypt freely, the cucumbers, and the melons, and the leeks, and the onions, and the garlick. But now our soul is dried away. There is nothing at all, besides this manna before our eyes" (Num. 11. 5. 6): In later years, when the Jews received the Mosaic statutes, we find *scaleless fish* prohibited as food (Lev. 11. 9–12). The *Phœnicians* on the coasts of the Mediterranean sea, carried on an extensive business in fishing, and great quantities were shipped from their chief ports, Tyre and Sidon. They carried on an extensive commerce in fish with distant countries, and we are expressly informed in Nehemiah 13. 16 that Tyrian fish mongers established themselves in Jerusalem, and desecrated the Sabbath by selling their commodities on that day to the people of Judah. That Jerusalem had a regular fish market, is implied in 2 Chron. 33. 14, and Neh. 3. 3. Fish are very abundant in all the waters of Palestine, except in the Dead

Sea, the waters of which are so salty that nothing can live in
it. The current of the river Jordan being very rapid, large
numbers of fish are swept into this Sea, where they soon perish,
and upon rising to the surface become the food of birds. The
sea of Galilee, (also called Gennesaret, Matt. 14. 34, &c.,
Tibterias, Jno. 21. 1), is especially noted for its fish. It is a
lake thirteen miles in length, by six in width.

Capernaum, which was once the home of Jesus (Matt. 4. 13),
and eight other cities stood upon its shores. It was on the
beautiful shores of this "Sea," that our Saviour spent most of
His days while upon earth, and here some of His mightiest
works were performed (Matt. 8. 5; 11. 23, &c.) On the shores
of this lake He selected His first disciples from the humble
fishermen, while they were actually engaged in their work.

After Peter's mournful confession that they had "toiled all
night," and had "taken nothing," they obeyed the Saviour's
direction to "launch out into the deep and let down your nets
for a draught" (Luke 5. 4), with the result of a miraculous
draught of fishes, thus proving to them indubitably His Divin-
ity, so that Simon Peter "fell down at Jesus' knees, saying,
depart from me; for I am a sinful man, O Lord" (verse 8).
At the same time, giving them a certain assurance of their
future success in the work of soul saving, by saying, "Fear not;
from henceforth thou shalt catch men" (verse 11). While the
day of Salvation lasts, the Gospel net will continue to be
thrown into the great sea of humanity, and draw on to the
"Rock of Ages," such as shall be saved (Matt. 13. 47).

Several modes of catching fish are mentioned in the Scriptures. With nets (Luke 5. 2; Ezek. 26. 5; Jno. 21. 6); with
hooks (Job 41. 1; Isa. 18. 8). Our Saviour told Peter to "go
to the sea and cast an hook, and take up the fish that first
cometh up, and when thou hast opened his mouth, thou shalt
find a piece of money, take that and give unto them for me and
thee" (Matt. 17. 27). This was "tribute money," which was a
Stater, a Roman silver coin of the value of four drachmas,
(about seventy cents) which was exactly the amount needed to
pay for both. For remarks on Jonah 1. 17, respecting the
"great fish," see article *whale* in zöology.

REPTILES.

SERPENTS. In ancient times serpents were, in many respects, regarded in a very different light from that in which they are regarded at the present day, serpents and crocodiles were objects of veneration and worship by most Pagan nations, and it is said some of the savage tribes of Africa regard the deadliest serpents with great veneration, and instead of destroying them, permit them to come into their huts and even feed them. The serpent figures prominently in the religion and mythology of most ancient nations, *Opiolitry* (serpent and dragon worship) was very generally practised. It was especially prevalent among the Egyptians, as is quite evident from the various representations of the serpent as an object of worship on the most ancient monuments. That the Babylonians and Assyrians practised it, is evident from the story of "Bel and the Dragon" in the Apocrypha.

To serpents was attributed a great longevity, and hence a serpent in the form of a ring became the emblem of eternity. The ancients also held that some serpents have great power over diseases, and from this superstitious notion, doubtless originated the emblem of health, which is a rod with a serpent twined around it.

Of the class *Reptilia*, Palestine possesses perhaps a greater number of species than any other region in proportion to its area. This, of course, is owing to its varied geographical features and wide range of temperature. Of the vast number of reptiles found, there are, however, comparatively few referred to in the Scriptures.

Of the order *Ophidia* (serpents), there are a great many genera, and species. Eighteen species were found by the party of Prof. Tristram during a short stay in the country, and it is probable that many more were not discovered by them. It therefore follows that on account of the great number of species, and the indefinite character of the Scripture references to them, a concise account of them cannot be given.

ADDER.

The genus *Vipera* (Viper) is well represented in Palestine. This genus embraces about one-fifth of all the serpents enumerated by naturalists. There are several species found in Palestine.

The viper may be briefly described as a very venomous serpent. The upper jaw is toothless but provided with two moveable fangs in front; connected with them are glands containing the poisonous secretion. Of over twenty species of this dangerous reptile may be mentioned, the *Vipera communis*, which is found in every country in Europe, except Ireland and the extreme North. The *Adder* of Scripture belongs to the genus *Vipera*, or viper family. Although the adder is several times mentioned in Scriptures, we cannot tell precisely what species is referred to.

PUFF ADDER (*Clotho Arietans*), found in the East, is from four to five feet long, and very thick in proportion to its length. It has a broad, flat head, and when irritated greatly distends the upper part of its body. The bite of the adder is very deadly (Psa. 140. 3; Prov. 23. 32).

Mention is made in Psa. 58. 4, of a "deaf" adder, "which will not hearken to the voice of the charmer." While there is a species of adder popularly supposed to be deaf, yet it has been shown that they are just as susceptible to the arts of the serpent charmer as any other. The above passage is evidently figurative.

Several species of the genus *Naja*, or hooded snakes, occur in the East. The best known are *Naja Haje* and *Naja tripudans*. These serpents have the power of puffing out to large dimensions the sides of the neck and head, which gives them the hooded appearance.

In Isa. 30. 6, we read of "*flying serpents.*" There is no such reptile known to naturalists at the present day, although they are mentioned in the mythology of the ancients, and represented on Egyptian monuments, as having the wings of a bird. Their existence is believed by the best authorities to be mythical, and the above passage is supposed to refer to serpents which have the power of swinging themselves a considerable distance, and dart upon their prey. A number of species are known to do this.

A serpent supposed to be several times alluded to in the Bible is the horned viper. The *Cerastis* has a broad, depressed, heart-shaped head, with a spinous or horn-like growth above each eye. Hence its name. Its bite almost invariably proves fatal in a short time. Although rare, it has been met with by travelers in Palestine.

Asp. The asp is mentioned as a deadly reptile in Deut. 32. 33; Job 20. 16; Isa. 11. 8, and Rom. 3. 13. Of this reptile nothing is certainly known, except that it was very poisonous. Some authors are certain that it is the *Naja haja*, while others with equally good reasons say it was a species of *Cerastis*. The asp of naturalists is a small and very venomous reptile called *Vipera Aspis*, found in some parts of Europe. The beautiful, but licentious *Cleopatra*, queen of Egypt, committed suicide by allowing an asp to bite her (B. C. 30).

Cockatrice. This creature is mentioned in Isa. 11. 8; 14. 29; 59. 5, and Jer. 8. 17. The cockatrice considered historically is wholly fabulous, since no creature answering to its description by ancient writers is known to exist. It was represented as being hatched by a toad from the egg of an extremely old cock. This and many other things concerning it, equally absurd, proves it mythical character.

Instead of *Cockatrice*, Luther gives the *Basilisk*, another fabulous creature, as the rendering. This creature is described by ancient Greek and Roman writers as a serpent of the most deadly character. The very breath of which was fatal to life. Later authors make it a creature of a lizard-like appearance, and having eight legs. It was said to be so poisonous that not only its breath, but its very look, was fatal. Pope Leo IV. (died A. D. 855), is said to have delivered Rome from one of these terrible monsters, the breath of which caused a pestilence. It was supposed to be the king of reptiles. Its name is the Greek diminutive for king (*Basiliscus*). After this fabled monster, Zöologists have founded the genus *Basiliscus*, an order of Saurian reptiles. The general supposition is that a species of *adder* is intended by the word translated *Cockatrice* and *Basilisk*.

In closing this article on the serpents of the Bible, it seems proper to add that the attempts of naturalists to identify the "*fiery serpents*" of the wilderness (see Num. 21. 6), with known species of serpents may be considered failures.

Several species of *Najas* occur in the regions traversed by the Jews, but there is no proof whatever that this serpent, or the *Cerastes* was identical the reptile.

The wonderful art of charming serpents was well known, and practised by the ancients (Psa. 58. 5; Jer. 8. 17).

DRAGON. Of this creature, quite often mentioned in Script· ure, but little can be said with accuracy. It seems to be a general term denoting any great monster, whether on land or in the deep. In Ezek. 29. 3, where Pharaoh is compared to a dragon, the reference is clearly to the *Crocodile.* "The great dragon that lieth in the midst of his rivers, which hath said, my river is mine own, I have made it for myself." The river meant is the Nile in which the crocodile abounds. In Isa. 13. 22; Jer. 9. 11; Isa. 34. 13, where the creature is represented as frequenting ruins and desolated places, we may suppose that any Saurian reptile inhabiting such places is intended. In Job 30. 29, it is supposed the *Jackal* is meant. In other places it doubtless refers to monster serpents, such as the *Boa* and *Python.* In Rev. 12. it is emblematic of *Satan.*

LIZARD (*Genus Lacerta*) (Lev. 11. 30). A genus of Saurian reptiles found in all temperate and tropical regions of the globe. There are hundreds of widely differing species of Lacerta. The *Chirotes* have no hind feet. In the bipes, the fore limbs are wanting, while the *Augius* and *Pseudopus* have no limbs at all, but are easily distinguished from true serpents by their structure. The species commonly known as lizards, have four limbs, having five toes on each foot. The body is generally slender, while the color of most species is variegated and very brilliant. Although generally detested by man, they are a harmless creature. Over twenty species have been found by naturalists in Palestine.

FROG. The *Batrachia* (frogs, toads, &c.) are not very well represented in the East, only a few species having been described by naturalists. In Egypt along the Nile is found the speckled grey frog, *Rana punctata.* Some authorities give only one species of frog as occurring in Palestine, the green trog — *Rana Esculenta.* Recent travelers, however, enumerate one or two more species, and one species of toad.

The frog is only mentioned in connection with the second plague in Egypt (Ex. 8.; Psa. 78. 45; and 105. 30). Only once in the New Testament, symbolically, (Rev. 16. 13.)

TORTOISE. Hebrew scholars are almost unanimous in the opinion that the word " *Tzab* " in Lev. 11. 29, rendered *tortoise* in the English, and *Kröte* (toads) in the German versions, refers to a species of large *lizard*. Perhaps, *Lacerta Nilotica* or Nile lizard, which is also met with in Palestine, and quite distinct from the species of the following verse (verse 30). The order *Chelonia* (tortoise, turtles, &c.) however, is well represented in Palestine, and a number of species have been described by naturalists.

CHAMELEON. A genus of Saurian reptiles which have a remarkable power of changing their color at will. There are only a few species of this interesting reptile, all natives of warm latitudes. It resembles a lizard in shape, but has a crest or ridge on its back. Its neck is so short that it cannot turn its head, but this defect is compensated by its eyes, which are large and move independently of each other. Owing to the peculiar structure of its feet it climbs trees with great ease. It subsists upon insects which it catches by darting out its long tongue with unerring accuracy. The tongue being covered with a viscous saliva the insects adhere to it.

The only species found in Palestine is *Chameleon Africanus*.

The Chameleon was classed with the unclean animals (Lev. 11. 30).

ORNITHOLOGY. (BIRDS.)

BITTERN (*Botaurus*) (Isa. 14. 23; 34. 11; Zeph. 2. 14). Much
difference of opinion exists among authorities as to the meaning
of the Hebrew word "*Kippod*," rendered "*Bittern*" in the
English, and "*Porcupine*" (*Igel*) in the German versions. (For
a further account, see "Hedgehog.")

BITTERN.

The Bittern (*Botaurus*) is an aquatic bird, belonging to a
sub-genus of the *heron* family. There are a number of species,
two of which are found in America. Generally speaking it is
a solitary bird, frequenting old ruins, and low, damp marshy
places. The species referred to in Scripture is probably
Botaurus Stellata, which is found in Europe and Western
Asia. It is a large bird of a dull yellow plumage, mottled with
black.

CRANE (*Grus*) (Isa. 38. 14; Jer. 8. 7). The crane is a large aquatic bird, and like all waders has long legs and a long bill. They differ from *storks* and heron chiefly in having the hind toe placed higher on the leg than the front ones. It is about four feet high when standing erect. It is quite probable that the "Crane" of Scripture is not the true crane, but a bird belonging to the same family now known by the name of

CRANE.

Anthropoides. It is the *Grus Virgo* of older naturalists. This bird (*Anthropoides Virgo*) is migratory (Jer. 8. 7), visiting Lake Tiberius yearly. It is about three feet high. Its general color is bluish grey, but its cheeks, throat, breast, and tips of the long hinder feathers are black. It has also a tuft of delicate white feathers behind each ear.

CORMORANT (Lev. 11. 17; Deut. 14. 17). A water fowl of the genus *Phalocrocoryx*, of which there are many species, found along the sea coast of almost every country. They are a very large and heavy bird, webb-footed, and live entirely on the

products of the sea. The prevailing color is black. The word
"Salach" rendered "Cormorant," seems not to be well understood by scholars. Luther always renders it "Swan." Prof.
C. H. Smith, an eminent authority on Bible Natural History,
holds that the *Tern* or large sea swallow, (*Sterna Caspia*)
accords best with the radical meaning of the Hebrew word.

CUCKOO (*Cuculus*) (Lev. 11. 16; Deut. 14. 15). There are
many species of this bird found in almost every quarter of the
globe, except the coldest. In a general way the *Cuckoo* may
be described as a good-sized bird. Its beak is compressed and
slightly ridged. Its wings are long, tail long and rounded. Its
color variegated.

Several species of
the Cuckoo do not
rear their own young,
but deposit their eggs
with their bill in the
nests of other birds,
where they are hatch-
ed and cared for by
foster mothers. This,
however, is not the
case with the Ameri-

CUCKOO.

can species (*Cuculus Americani*).

COCK (*Gallus domesticus*). It is regarded as strange that
there is no reference made to this domestic fowl in the Old
Testament. From its silence, it is presumed that the chicken
was not well known to the Jews during the early period of their
history. In the New Testament, however, the references are of
such a character as to indicate that it was well known to the
Jews at the time of Christ; and was tolerated as a clean bird.
The various "breeds" known to us are all varieties of one
species (*Gallus domesticus*). Our Saviour draws a beautiful
illustration from a hen hovering over her brood (Luke 13. 34).
Crowing of the cock (Mark 13. 35; Matt. 26. 34 and 74;
Mark 14. 30).

DOVE (*Columba*). Also known as the *Pigeon*. The species
of this bird are so numerous and widely differing, that a

description cannot be attempted here. They are found in all temperate quarters of the globe. As a bird of flight, it is, perhaps, unexcelled, while some species, especially those found in the tropics have plumage of great beauty. Viewed scripturally, the dove is a very exalted bird. By Divine appointment it became the emblem of the Holy Ghost (Matt. 3. 16; Mark 1. 10; Luke 3. 22; John 1. 32). Sent out of the ark by Noah (Gen. 8. 7). Used in sacrifice by Abraham (Gen. 15. 9). Under the Levitical law generally by the poor (Lev. 1. 14; 5. 17, &c. See also Luke 2. 24; John 2. 14).

Its powers of flight referred to (Psa. 55. 6; 68. 13). Its plaintive voice (Isa. 38. 14; 59. 10). The beauty of its eyes (Sol. Song 1. 15; 5. 12). Its innocence (Matt. 10. 16). Domesticated (Isa. 60. 8). Used figuratively (Song 2. 14; 5. 2; 6. 9). Carrier Pigeons, with messages attached to them, are portrayed on Egyptian monuments of great antiquity.

DOVE.

EAGLE (*Aquila*). The eagle family comprises many species, differing very considerably in size, plumage and habits. Some are land birds, while others depend entirely on the water for subsistence. The eagle is distinguished for its size, courage, arms for attack, and extraordinary powers of flight. The finest species is the *Golden Eagle* (*Aquila Chrysætos*), which frequents all temperate regions of the globe. A large species quite common in Palestine is *Aquila Heliaca*. The eagle is a bird of prey, and was, therefore, unclean (Lev. 11. 13) They delight in inaccessible cliffs and lofty elevations where they rear their young (Jer. 14. 16). Beautiful illustrations are drawn from the eagle's powers of flight (Deut. 28. 49; Prov. 23. 5; Jer. 48. 40; Hab. 1. 8; Ex. 19. 14; Psa. 103. 5; Isa. 40. 31). In Matt. 24. 28 and Luke 17. 37, the Saviour doubtless mentions it as a symbol of the Roman army, the Insignia of which was the eagle.

GIER EAGLE (Hebrew, *Racham*) (Lev. 11. 18; Deut. 14. 17). There is no such bird now known to naturalists as the "*Gier Eagle.*" The best authorities hold that the bird intended is the white vulture (*Neophron Percuopteris*). This is the smallest species of the vulture family, being but little heavier than the raven. Its bill is long and slender, and terminates with a hook. Its head and throat is devoid of feathers. Its general color is white. It is the vilest of birds, living exclusively on carrion. It is very plentiful in the hot countries of the East. In Egypt it is called "Pharaoh's Chicken," and is protected by law, because of its usefulness as a scavenger.

GLEDE (Deut. 14. 13). Great uncertainty exists as to the distinction between "Glede" and "Kite," both words occurring in this passage, and in English meaning the same thing; *Glede* being an antiquated expression for *Kite.* No uniformity exists in the various European versions. Luther renders it "*Taucher*" (Diver), a name applied to several species of aquatic birds. Several eminent authorities hold that the original refers to a species of vulture.

HAWK (*Falco*) (Lev. 11. 16; 14. 15; Job 39. 26). This bird belongs to the same family as the eagle (*Aquila*), but differs from that genus in several particulars. The *Falcon* tribe is represented in all temperate regions of the globe. All hawks are birds of prey, and consequently have strong bills and claws, wherewith to seize and devour their prey. There are many species, differing considerably in form, color and habits. Some are exclusively land birds, while others, although they are not aquatic, yet depend upon the products of the water for subsistence. Several species can be domesticated and trained to capture small game, such as birds and small animals.

HERON (*Ardea*) (Lev. 11. 19; Deut. 14. 18). The *Heron* is an aquatic bird of large size, living entirely upon the products of the water, such as frogs, fish, and other small objects. The species most common in Europe is *Ardea Cinerea,* which is about three feet in length, and of a beautiful grey color, except the quill feathers, which are black.

A species of rare beauty is the white heron (*Ardea Alba*), which is an occasional visitor in Europe from the far East. It is a very large bird. Its plumage is loose and flowing, and perfectly white. The species most common in Palestine is *Ardea Herodius*, which has a beautiful mottled neck and breast, and several long feathers on the top of the head. Like all waders the heron has long legs and bill.

HERON.

KITE (*Milvus*) (Lev. 11. 14; Deut. 14. 13). Also known as the *Glede.* The kite is a bird of prey and belongs to a sub-genus of the Falcon family. It is especially noted for the towering height to which it soars, and the great velocity with which it descends upon its prey. The common species, *Milvus Vulgaris,* is about two feet in length from the tip of its bill to the end of its tail. Its plumage is of a brown color. Its wings are long, and extend to the end of the tail, which is forked like that of the swallow. The species probably best known to the Jews, was the black kite (*Milvus Ater*).

LAPWING (Lev. 11. 19; Deut 14. 18), (German translation *Wiedehopf*). It is agreed by nearly all authorities that the Hebrew word, "dukiphath," meaning "*double-crest,*" does not apply to the *Lapwing* (*Vanellus*), but to the *Hoopœ*, (*Upupa*). In this, most versions agree, (Wiedehopf is the German for *Hoopœ*). The Hoopœ is a very peculiar bird in appearance, voice and actions. It is about the size of the pigeon. Its plumage is of a russet color, mixed with white and black. Its wings and tail are crossed by white bars. On its head are two

KITE.

parallel rows of long feathers tipped with white and black. This bird has always been associated with the magic and super- stition of the East. Owing to its peculiar habit of slowly depressing its head until its bill touches the ground, and at the same time giving its head a peculiar motion, it was supposed to dictate the presence of hid treasures, and reveal secrets. Its head was used by the magicians and charmers, with which to perform their incantations. Its name is derived from its peculiar cry, which resembles the rapid articulation of the syllable " Hoop."

Night Hawk, Hebrew, *Tuchmas.* The "Night Hawk," properly speaking, is no hawk at all, but is that genus of birds known to science as the *Caprimulgus.*[*] Of this peculiar bird there are several species found in Europe and the countries of the Mediterranean basin. It is an insectiverous bird, about the size of the raven. Color variegated, eyes large, and adapted for night vision. Tail forked, like that of the swallow. Many superstitions were connected with this bird by the ancients.

LAPWING.

It was an ill omen to see one after eventide. They were supposed to suck the udders of goats at night, and hence arose the name " *Caprimulgus,*" (literally *"goatsucker,"*) by which they are generally designated.

Ospray (*Pandion Haliaëtus*) (Lev. 11. 13; Deut. 14. 12). A sub-genus of the Falcon family. Also called *Fish Hawk* and Fishing Eagle. It is found in all regions of the globe,

---

* Goatsucker. 6

except the coldest. It is almost two feet in length, color dark-brown, mixed with black, grey and white. The bill is short, rounded, and very strong. Wings long, and extending behind the tail. It lives exclusively on fish.

OSTRICH (*Struthio Camelus*) (Job. 39. 13: Lam. 4. 3). "*Taanah,*" (*Ostrich*), occurs in several other places in the Bible, but is supposed to be mistranslated in the English version, notably in Lev. 11. 16. and Deut. 14. 15, where it is rendered "Owl." The German and some other versions render it ostrich in the last quoted passages. The ostrich is

OSTRICH.

the largest bird known to man, being from seven to eight feet in height, and weighing sometimes 300 pounds. Its neck is long, head rather small, and both neck and head destitute of feathers, but covered with light down. Its legs are long and powerful. The foot has but two toes; the inner one is the largest (about seven inches in length), and terminates with a hoof-like claw. The plumage of the male is black, and of the female dark grey. The "plumes" so much prized are obtained from the wings, which produce about twenty on each. The wings, which are naked underneath, are short, and unsuited for flight, but are used by the bird to assist in running. The speed of the ostrich is far greater than that of the fleetest horse, but its capture is made possible by its habit of running in a curved course. It is a vegetable feeder, but has a curious habit of swallowing various indigestible substances, such as stones, shells, sticks, and even metals. The female ostrich lays her eggs in the hot sand, where they are hatched mainly by solar heat, but the birds do not utterly forsake their nests, as is generally supposed. A full-grown ostrich can easily carry two ordinary sized men on its back.

OSSIFRAGE (*Aquila Ossifrage*). Better known as *Gypætes Barbatus* (Lev. 11. 13; Deut. 14. 12). This is a powerful bird of prey, classed by some in the eagle family. It is over four feet in length from the tip of its bill to the end of its tail, while its wings have an expanse of about ten feet. Its head and neck are covered with narrow feathers of a white color, while under its bill there is a tuft of bristly hair. Its general color is black and brown, with white stripes on its wings and neck. This bird is also found in Eastern Europe, where it is said not to seize large prey in a live state, but to pursue game, such as goat, chamois, deer, &c., in the mountains, where in various ways it causes its prey to fall over the steep cliffs, where they are killed by the fall, and afterward devoured by the bird.

OWL (*Strix*). A genus of birds of the family *Strigidæ*, or nocturnal birds of prey. This genus of birds differs widely from all others. They are especially distinguished by their erect posture, large head and eyes. The eyes of all the species are surrounded by a disc of feathers, which radiate outward. The bill is very short, and sharply curved. Both bill and claws are adapted for seizing and destroying its prey. There are more than twenty species of owls, differing very much in size and appearance. The smallest kind is about seven inches in length, while the largest species, such as the Snowy Owl, (*Strix Nyctea*), and Eagle Owl (*Strix Bubo*) are powerful birds. The latter is but little inferior to the golden eagle, and does not hesitate to seize hares, rabbits, young deer, &c. The common barn or screech owl (*Strix Flammea*) is believed by naturalists to be the most widely diffused bird on the globe, being found in almost every country inhabited by man.

The owl is a strictly nocturnal and solitary bird. It delights in old ruins and dark, dismal places (Isa 13. 21; 34. 13; Jer. 50. 39). The voice of some is very doleful (Mic. 1. 8). It was unclean to the Jews (Lev. 11. 16; Deut. 14. 15).

Several species of owls are enumerated in the Bible.

"*Little Owl*" (Lev. 11. 17; Deut. 14. 16). This is supposed to be the common barn owl (*Strix flammea*) already referred to.

"*Great Owl*" (Lev. 11. 17; Deut. 14. 16). Believed by good authorities to be the *eagle owl* (*Strix Bubo*) already mentioned.

"*Screech Owl*" (Isa. 34. 14). The original is correctly rendered "night monster" in the margin of the English version. Luther renders it "*Kobold*," the equivalent of *Spectre*, or hobgoblin. It is a question, therefore, whether this is not a figurative expression.

Owls have in all ages been regarded as a bird of ill-omen, and harbingers of misfortune.

PARTRIDGE (1 Sam. 26. 20; Jer. 17. 11). Partridge is a somewhat general term denoting many widely differing species of the genus *Tetrao* or *Grouse* family; more than half a dozen species occur in Palestine alone, the most important of which we will mention. The *Greek* partridge (*Tetrao Saxatilis*) is very abundant. Plumage variegated, having its flanks and thighs barred with white, black and fawn colored feathers. The *Red-legged* partridge (*Tetrao (Perdix) Rufa*). A smaller species than the above. Included in the partridge family is the genus *Pterocles*. Of this very beautiful bird there are, perhaps, four species, native to Palestine. The *Pin-tailed* Grouse (*Pterocles Alchata*). The *Sand Grouse* (*Pterocles Arenarius*). The Arabian Grouse (*Pterocles Arabicus* and *Pterocles Senegalensis*). This genus of birds is distinguished by its long wings.

The *Francolin*, or tree partridge, is another genus of partridges found in Palestine. Of this genus the species best known is *Francolinus Vulgaris*, which is a magnificent bird.

The "Quail" species will be considered in its appropriate place.

PIGEON (See Dove). Pigeons were used in sacrifice chiefly by the poor (Lev. 12. 6; Lev. 5. 7; Luke 2. 24, &c.).

PEACOCK (*Pavo*) (1 Kings 10. 22; 2 Chron. 9. 21). Of this genus of birds but two species are known. The common *Pea* fowl (*Pavo Cristatus*) and the Japan Peacock (*Pavo Japonensis*) discovered some years ago. The Peacock is a gallinaceous bird, distinguished for its fine plumage. It is a bird of large size. Its head is crested. The bill is slightly arched. Its legs are rather long and armed with spurs. Its wings are rather short, and not adapted for prolonged flight. The chief peculiarity of the peacock is its magnificent tail feathers, often six feet in length, which it has the power of erecting in the form of a semicircular disc.

PEACOCK.

From the above references to the peacock, it is evident that it was not native to Palestine, but came from a distant country. The voyage to Tarshish and return required a period of three years. Much uncertainty exists as to the identity of "Tarshish." (See "Ape.")

PELICAN (*Pelicanus*) (Lev. 11. 18; Deut. 14. 17; Psa. 102. 6). The Pelican is a web-footed, aquatic bird, about the size of the goose. Its bill is very long and broad. The upper mandible is terminated by a strong hook which curves over the lower one. Under its long bill is a pouch of loose skin, in which the bird stores up its surplus food, chiefly fish, to eat at its leisure, or carry to its young. Its general color is white,

with the quill feathers black. It is said also to carry water as
well as food in its pouch to its young.

Pelicanus Onocrotalus occurs in the East and the species
with which the Jews were probably acquainted.

PELICAN.

QUAIL (*Coternix*) (Ex. 16. 13; Num. 11. 31, 32; Psa. 105.
40). This genus of birds belongs to the grouse family. It
differs from the true partridge in having a more slender bill,.

shorter tail, longer wings, and in being destitute of spurs. The quail is about seven inches in length. Its color is brown of various shades. The throat is generally white. The quail is a migratory bird, and is generally seen in flocks.

RAVEN (*Corvus Corax*). The Raven is a species of crow, from which it is, however, easily distinguished in being much larger. It also differs greatly in its habits. The raven is of a jet black color, about two feet in length from the tip of its bill to the end of its tail. The wings are long, having over four feet expanse. It is a thoroughly omniverous bird, eating almost anything that comes in its way, even destroying hares and lambs. "The raven after his kind," meaning the entire crow family, was therefore unclean (Lev. 11. 15; Deut. 14. 14). It is not a gregarious bird, but is nearly always seen singly or in pairs. The raven is the first bird mentioned in the Bible (Gen. 8. 7). A raven miraculously fed the Prophet Elijah every morning and evening (1 Kings 17. 4). Mentioned by the Saviour (Luke 12. 24). (See also Job. 38. 41; Psa. 147. 9; Prov. 30. 17; Cant. 5. 11, and Isa. 34. 11).

SWALLOW (*Hirundo*) (Psa. 84. 3; Prov. 26. 2. Isa. 38. 14; Jer. 8. 7). A genus of birds found in almost all countries, and containing a large number of species. All the birds of the swallow family are rather small. The common chimney swallow (*Hirundo Rustica*) is found in most parts of the world. It is hardly more than eight inches in length. Color bluish-black, chestnut and white. Tail long and forked. It builds its nest of mud. The window swallow or house martin (*Hirundo Urbica*) resembles the chimney swallow, but is quite different in its habits. The "swifts," or black martins are now generally considered by naturalists as forming a distinct genus of birds, comprising several species. This genus, now called *Cypselus*, was formerly *Hirundo Assus*. It has a forked tail like the swallow. Its wings are curved like a sickle. All the above birds are found in Palestine, and several other species of the same family, of which a description seems unnecessary. Several species of swallows build nests that are *edible*, and are an important article of commerce in China.

SPARROW. The Hebrew word *"tizppor"* seems to be a general term to denote any bird of the sparrow family which comprises many species. The word is rendered merely "bird" and "fowl" in most cases in the English version, except in Psa. 102. 7, and Psa. 84. 3, where it is rendered *Sparrow.*
Sparrow also occurs in Matt. 10. 29 and Luke 12. 6, where the Saviour says, "Are not two sparrows sold for a farthing?" From this it is inferred that they were used as food, as they still are at the present day. There are many species of sparrows. The common, or house sparrow (*Passer Domesticus*) is very abundant in the East. It is a small bird with grey and black plumage. Quill and tail feathers brown. It loves the habitation of man, and is so familiar that their great numbers are considered a serious evil in some places.

STORK (*Ciconia*). A genus of birds allied to herons, bitterns and other large waders (*Grallators*). There are three principal species of this bird. The white stork (*Ciconia Alba*). The black stork (*Ciconia Niger*) and the American stork (*Ciconia Maquari*). The white stork is about three and a half feet in length. The plumage is pure white, except the wings, which are partly black. Its bill and legs are very long, and of a bright scarlet color. When standing, it arches its neck very gracefully. The feet of storks are partly webbed. The black and American species are smaller than the white.

Storks have been regarded with great favor by man from time immemorial. It was regarded as a sacred bird by nearly all Oriental nations. To this day it is seldom molested even in Europe, where it is protected by law in some countries. In Holland, large, open boxes are sometimes placed on the roofs of houses for the stork to breed in, and it is considered a very fortunate circumstance if a stork selects one as her nest.

Storks are proverbially affectionate to each other, and especially to their young, defending them with their lives, and in most cases perish with them, if they are attacked. Young storks are also said to feed and care for their aged parents, but this is denied by most naturalists.

The stork is a migratory bird, and although it has no voice, yet by some strange instinct great numbers suddenly gather at one place, and then amid great clatter of bills and flutter of wings they start off (always at night) for the distant south; when once started, it is supposed they do not stop to rest until their destination is reached. It is held by eminent authors that they make the long journey from the lower Rhine (Germany, Holland, &c.) to Western Asia, and the Nile Delta in Africa, without stopping. Its migrations are noticed in Jer. 8. 7, "Yea the stork knows her appointed season." To the Jews it was unclean (Lev. 11. 19; Deut. 14. 18). (See also Psa. 104. 17, and margin of Job 39. 13.)

SWAN (*Cygnus*) (Lev. 11. 18; Deut. 14. 16). This genus of birds embraces several species of very large, web-footed, aquatic fowls. The best known is the common domesticated swan (*Cygnus Olor*). Its color is pure white. Length about five feet, weight when grown, about thirty pounds. Neck very long and arched. A small black knob is found at the base of the upper mandible of the bill. Swans live to a very great age. A black swan (*Cygnus Atratus*) is found in Australia. Swan occurs only in Lev. 11. 18, and Deut. 14. 16, as an unclean bird, whereas, being considered a vegetable feeder, some perplexity has been felt concerning its classification among the unclean birds, and especially so since the swan and its allied genus *Anser* (the goose) has in all countries been highly esteemed as food. Serious doubts are, therefore, entertained by critics as to the correctness of the translation. Recourse must, therefore, be had to original sources.

Both the Greek *Septuagint* and Latin *Vulgate* render the original Hebrew *Tinshemeth* by a water fowl, but not of corresponding natures. The Latin gives the equivalent of *Swan*, but the Greek gives "*Porphyrion*" ("A red colored water bird," Liddell & Scott's Standard Greek Dictionary). Scholars who take the vulgate as their authority, of course, maintain the correctness of the English version. In several respects, however, the *Septuagint* rendering is the preferable one. The *Porphyrion* is a medium sized bird. Its color is a dark and brilliant indigo, but its head, neck and sides of a beautiful turquoise. It has a hard crimson shield on the forehead. Its legs and toes are very long and flesh-colored. Since the originals differ in the rendering of the Hebrew, the reader must judge for himself as to which is the preferable rendering. The German instead of "swan" gives "bat"(*fledermaus*) and gives "swan" for the English "Cormorant" in Lev. 11. 17.

TURTLE-DOVE (*Turtur*). A sub-genus of birds of the Pigeon (*Colombo*) family. They differ from the true dove in being more slender in form, its bill is also more slender and slightly bent. Its wings are longer and more pointed. The tail is long and rounded. The color varies much in the different species. It is a mirgatory bird. "The *turtle*, and the crane, and the swallow, observe the time of their coming" (Jer. 8. 7). Its soft cooing voice was a welcome harbinger of Spring. "For lo the Winter is past, the rain is over and gone. The flowers appear on the earth. The time of singing of birds is come, and the voice of the *turtle* is heard in our land" (Sol. Song 2. 12).

It was a clean bird used for sacrifice (Lev. 5. 7; Num. 6. 10, &c.). It was offered by Mary, the mother of our Saviour, for that purpose (Luke 2. 24).

VULTURE (*Vultur*). A bird mentioned in Lev. 11. 14, and Deut. 14. 13, as unclean. Vultures form a genus of rapacious birds, embracing many species which differ widely in size, color and habit. Thus the sub-genus *Gypætos* (see "*Ossifrage*") classed by some naturalists in the vulture family, more closely resembles the genus *Aquila* (eagle). For a description of the white vulture (*Neophron Percnopteris*) see article on "Gier Eagle."

All the species of vultures are extremely filthy and disgusting birds, living almost entirely upon carrion, which by some strange instinct, or powerful scent, they soon discover, though far distant. The head and neck of most species are bare, or covered with slight down, while at the base of the neck there is a ruff or collar of feathers. The bill of all vultures is hooked. Their wings are long, and their powers of flight wonderful. Their power of vision, too, is extraordinary. It is alluded to in Job 28. 7: "There is a path which * * * the vulture's eye hath not seen." The whole tribe seem to be designed by nature as the scavengers of the world, and thus prevent pestilence and death to mankind in the hot countries in which they are chiefly found. It is on this account that some species are protected by law, (even in the United States). Several species are found in Palestine, some of which we have already described; we will yet mention the monster *Griffon* vulture (*Vultur Fulvus*) found in Eastern Europe, and also an inhabitant of Asia and Africa. It is more than four feet in length. Its general color is yellowish brown. Its quill and tail feathers are of a darker color. The down on the head and neck and also the ruff is white. Vultures are figured on ancient Egyptian monuments as following armies and hovering over battlefields in expectation of devouring the dead (See Isa. 34. 15).

ENTOMOLOGY. (INSECTS.)

Although insect life is very abundant in Palestine, the number of species mentioned in the Bible is comparatively small,
while naturalists have described and classified tens of thou-
sands of species of insects, of which over a thousand species
are found in Palestine. Yet the number enumerated in the
Bible is less than twenty.

The science of Entomology embraces seven orders of insects,
which we will enumerate and classify. The insects of the Bible
in their appropriate orders are

1. COLEOPTERA. Insects that have sheaths or shells that
cover the wings, such as beetles, &c. Over four hundred
species of beetles have been found by naturalists in Palestine;
we cannot say that any particular species is alluded to in the
Bible. True, the word "beetle" occurs in the English version
in Lev. 11. 21-23, yet it is very doubtful whether the rendering is correct. The original " *Chargol*" is left untranslated in
a number of European versions. The Septuagint and Vulgate
both have "*Ophiomachus*" (Serpent fighter). The general
opinion is that an insect of the *Locust* tribe is intended.
Should the reader desire to investigate the subject more closely,
he can refer to Kitto's Enc., article "Chargol," where he will
find the subject elaborated, and a species of Locust (*Truxalis
Nasutus*) figured as the insect probably intended. The
strongest argument, however, against a beetle being the insect
intended in Lev. 11. 21, is the fact that it was enumerated as

a proper article of food, while it is a well-known fact that neither Jews, nor any other enlightened nation, ever used beetles as food.

2. HEMPITERA. Half-winged insects, such as grasshoppers, locusts, crickets, &c. This order of insects is well represented in Palestine, upwards of forty species of locusts having been found there. There are nine different Hebrew words in the Bible, which are believed by good authorities to denote various species of locusts, not only the perfect insect, but also its *pupa* and *larva*.

It is impossible, however, to determine with accuracy to what particular species, or to what states of maturity in the insect, the various words refer; we will, therefore, give the various Hebrew words, and the manner in which they are rendered in the English version.

1) *Arbeh.* This word occurs in Ex. 10. 4, and in many other places in the Old Testament, and is always rendered *Locust* except in Job 39. 20; Judges 6. 5; 7. 12, and Jer. 46. 23. In the foregoing places it is rendered *grasshoppers.* The word seems to denote the locust in general.

2) *Gob.* Rendered locust in Isa. 33. 4, and *grasshopper* in Amos. 7. 1, and Nahum. 3. 17.

3) *Gazam.* Occurs in Joel 1. 4; 2. 25, and Amos 4. 9, and is rendered "*Palmer-worm.*" Palmerworms are no particular species of insects, but are the destructive *larva* of various winged insects, and in the above passages denotes the locust in its larval state.

4) *Chargab.* Rendered *grasshopper* in Lev. 11. 22; Num. 13. 33; Eccl. 12. 5, and Isa. 40. 22, and *locust* in 2 Chron. 7. 13.

5) *Chasil.* Occurs in 1 Kings 8. 37; 2 Chron. 6. 28; Psa. 78. 46; Isa. 33. 4; Joel 1. 4, and 2. 25, and is rendered *Caterpillar* in the English version. The *Septuagint* and *Vulgate* has it *Bruchos* (a Greek word denoting a wingless locust, Liddell & Scott's Greek Dictionary). The Latin authors understood the word to denote the *larval,* or *pupa* state of the locust, but the Greeks applied the word to the perfect insect -- well.

6) *Chargol.* Rendered *"beetle"* in Lev. 11. 22, but left untranslated in most other versions on account of the uncertainty of its meaning. The general agreement that it denotes some kind of locust has already been noticed.

7) *Yelek.* This word is rendered *Caterpillar* in Psa. 105. 34; Jer. 51. 14, and *Cankerworm* in Joel. 1. 4; 2. 25, and Nahum. 3. 15. The *Septuagint* and *Vulgate* in most instances renders it *Bruchos.* A wingless locust (See *Chasel*).

8) *Salam.* Rendered *"Bald Locust"* in Lev. 11. 22, but left untranslated in most versions. Several authorities are of the opinion that the species of locust intended is *Gryllus Eversor.*

9) *Tzelatzel* (Deut. 28. 42). Rendered locusts. Supposed to be the species *Gryllus Striadulus.*

Having considered the various Hebrew originals denoting the locust and its larva, a few additional remarks will be in place. As already remarked, locusts are very abundant in Palestine, the most destructive species of which is *Locusta Migratoria*, or the migratory locust.

Locusts are the most destructive insects known. They destroy all kinds of vegetation, not even sparing the tender bark of trees, and a country visited by them presents a scene of indescribable desolation and misery. They migrate in swarms, following the wind, and so great is their number that they darken the sun for several hours in their passage. Their visitation is considered a national calamity, and we can, therefore, understand how intolerable must have been the plague of locusts brought on the Egyptians (Ex. 10. 4). We also see the import of the figurative references to the locust, such as Psa. 109. 23; Joel 1. 4; Isa. 33. 4, and Rev. 9. 3. We have already seen (Lev. 11. 22), that locusts were permitted to the Jews as food. It is a fact that most Eastern nations use it for this purpose even at the present day, and they are said to be wholesome and nutritious, and very palatable. They are prepared in various ways; stewed in butter, or oil, dried and preserved like herring, or dried thoroughly and ground into meal. Hence the eating of locusts by John the Baptist is to be considered literally (Matt. 3. 4; Mark 1. 6).

In the order *Hemiptera* also belongs the *Coccus*, insects comprising several genera belonging to the sub-order *Homoptera*. Although these insects are nowhere mentioned in the English version, yet the dyes obtained from them are quite frequently mentioned. The *coccii* are parasite insects, adhering to plants, trees, &c., from which they can scarcely be distinguished sometimes. They suck the juices of plants, and some species are very troublesome. Several species, however, are very valuable on account of the beautiful dyes obtained from them. That best known to us is the *Cochineal*, (*Coccus Cacti*), which is found chiefly in Mexico, where it feeds on the cactus. Of the female bodies of this species is made *Carmine* and lake dyes. The species known to the Orientals is *Coccus Ilicis*, which produces the beautiful "crimson," so often mentioned in the Bible. This little insect is found upon the Holm oak, (evergreen). It is about the size of a pea, violet black in color, covered with a whitish powder. The males have wings, and do not thus adhere. So stationary are these insects that early naturalists were ignorant of their insect character, and supposed they were the product of the tree. Where "crimson" indicates this insect or the dye obtained from it, the Septuagint and Vulgate is specific, giving the name of the insect itself (*Kokkinon, Coccus, &c.*).

3. LEPIDOPTERA. Insects with scaly wings, such as the butterfly, moth, &c. Although almost three hundred species of the butteflies and moths have been found in Palestine, yet there is not a single allusion to any of them, except the *Tineidœ*, the *larva* of which are known as "*moths.*" The passages where it occurs are as follows: "They shall wax old as a garment. The *moth* shall eat them up" (Isa. 50. 9). "He consumeth as a garment that is *moth* eaten" (Job 13. 28). "Where *moth* and rust doth corrupt" (Matt. 6. 9). (Also Job 4. 19; Isa. 39. 11; Isa. 51. 8; Hos. 5. 12; Jas. 5. 2). Of the various kinds of this insect, we will mention only a few of the most important, *Tinea destructor*, a small insect of a buff color. Wings deflexed when at rest. Its *larva* (the mouth) is about a quarter of an inch long, with a few white hairs on its body. This is the most destructive of all moths,

Tinea Tapezana destroys woolen goods; *Tinea Pellionella* destroys furs; *Tinea Granella* destroys books and grain.

4. NEUROPTERA. Nerve-winged, or fiber-winged insects, such as the "Dragon fly," "May fly," "Trout fly," &c. Although this order of insects is well represented in Palestine, not a single species is mentioned in the Bible, and a description is, therefore, not necessary here.

5. HYMENOPTERA. Insects having four wings and a sting, such as the bee, wasp, hornet, &c. Of the hundreds of species of insects belonging to this order, and found in Palestine, only a few are mentioned in the Bible.

ANT (*Formica*). There is, perhaps, no insect in the world of greater interest to the student of nature, than the *ant*, and on some accounts it is the most remarkable of all. That its habits were observed, and known to the ancients, is evident from the following allusions to it in the Bible: "Go to the *ant*, thou sluggard, consider her ways and be wise, which having no guide, overseer, or ruler, provideth her meat in the Summer, and gathereth her food in the harvest" (Prov. 6. 6–8). "The ants are a people not strong, yet they prepare their meat in the Summer" (Prov. 30. 25). There are over a dozen species of ants found in Palestine, some an inch or more in length. A physical description of the various species cannot be attempted here, and we will confine ourselves to a notice of their chief peculiarities. All ants have extraordinary strength, and some species can carry twelve times their own weight. Had mankind strength proportionate to the ant, it follows that a man weighing 166⅔ pounds would be able to carry a weight of one ton. Their intelligence is something surprising, and far beyond ordinary insects. They live in communities like bees, and like them, are divided into three classes, viz.: *males, females* and *neuters*. The neuters greatly outnumber the males and females, and are themselves again divided into two classes, *soldiers* and *workers*, while it is literally true that they have no "guide, overseer or ruler," yet there is no mutiny or lack of discipline in the standing army. The workers are always busy. Every one finds something to do, and each one seems to have its alloted work, and pursuing its definite purpose. They bestow great care upon their eggs, or pupa. Some are carried out and given a sun "bath." Some are licked and manipulated, for what purpose is not definitely known. Some are arraigned in rows, or corded on piles, and, strange to say, at the precise time when the embryo is ripe, the egg is opened by an ant, and the young comes forth and soon gets to work. The males and females have wings, but the neuters, which are by far the most numerous, have none. Some species of ants are provided with a sting, while others defend themselves by discharging a very fetid fluid (*Formic acid*).

BEE (*Apis*). The bee family embraces many hundred species. The only one referred to in the Bible is the *honeybee* (*Apis Mellifica*), and our observations must be confined to it. The importance of the honey-bee to mankind is seen in the attention bestowed upon it. Eminent men, both in ancient and modern times, have made the bee the object of a life-long study. Hundreds of books have been written, and many magazines are published, illustrating its nature, habits, &c., while societies have been formed with a view of promoting a knowledge of bee culture. The natural history of bees is full of interest to the observer. Honey-bees live in communities, and are divided in three classes, *males, females* and *neuters*. There is but one female (the "queen") to each community, and she maintains her supremacy by destroying all rivals. She is generally attended by a sort of body guard, or retinue of from fifteen to twenty bees. During the warm season she lays the eggs in the cells provided for her by the workers. When the community becomes very large, and a part desire to start a new community, the old queen is carefully guarded to prevent her from killing the young queen, which is also cared for by the prospective emigrants. When she is sufficiently mature, the "swarming" takes place. Bees are found in all parts of the world where there are flowers enough to afford them subsistence. In some countries they make their home in the clefts of rocks, and hence the Psalmist's expression, "Honey out of the rock" (Psa. 81. 16). In some places bees are so numerous and fierce as to be a serious inconvenience to the inhabitants. The ancient historian, Pliny, informs us that in some parts of

7

Crete bees were so troublesome that the people were obliged to forsake their homes, and remove to more congenial quarters. This throws great light on the assertion of Moses, "And the *Amorites* which dwelt in that mountain, came out against you, and chased you as bees do" (Deut. 1. 44). Also the phrase, " They compassed me about like bees" (Psa. 118. 12). For a remarkable bee's nest see Judges 14. 8. Honey is mentioned about forty times in Scripture, and was no doubt very abundant in Palestine, so that it was literally "a land flowing with milk and honey " (Ex. 3. 8). In warm countries bees do not confine themselves to cavities in which to form their combs, but sometimes choose the forks or branches of trees. This fact may throw some light on 1 Sam. 14. 24–32.

HORNET (*Vespa Crabo*). An insect of the *Vespidæ*, or wasp family. Like bees and ants, they are divided into three classes, *males, females* and *neuters*, or *workers*. The males and neuters perish in the Winter, while the females survive to originate new communities in the Spring. Hornets are very voracious, and devour all kinds of small insects. They are also very fierce, and their sting is very painful. They were sent by Jehovah to extirpate the enemies of his people (Ex. 33. 28 ; Deut. 7. 20; Josh. 24. 12).

6. DIPTERA. This order embraces the insects having two wings, such as the *gnat, fly, musquito,* &c. To the insects of this order, which are small, the general term *"fly"* is sometimes applied. In a more restricted sense, the term fly is applied to the genus *Musca,* which embraces many species, such as the *house-fly, bow-fly, crane-fly, bot-fly,* &c. The fly, proper, is distinguished for the facility with which it walks on smooth surfaces, such as glass, &c., even in an inverted position, notwithstanding the observations of naturalists, the manner in which it is done is not clearly understood. Flies are provided with two large compound eyes, and some species have several additional ones on the top of the head. Nearly all dipterous insects have a proboscis, with which they sucks their food (blood, juices, &c.).

Two different Hebrew words are rendered "fly" in the English version. The first is *Aroob* (Ex. 8. 21. &c., Psa. 78. 45, and

Psa. 105. 31). These passages all relate to the plague of flies in Egypt. In the Septuagint the word is uniformly rendered *Kunomnia* (dog-fly). Luther renders it *"ungeziefer,"* a general term for vermin and noxious insects. The next word is " *Zebub*," and is found in Eccl. 10. 1, and Isa. 7. 18, (Sept. *Musca*, " fly," Luther concurs). This is a general term for fly in several Semitic languages, and hence *"Baal,"* god or lord and "*Zebub*," " fly." Literally *fly-god*, or " lord of the flies." A god of the Philistines (See 2 Kings 1. 2).

The only other dipterous insect mentioned in the Bible is the *gnat* (*culex*). The common *gnat* (*Culex pipens*) is also found in Palestine. Like the fly, it has a proboscis, with which it " bites" and through which it sucks its food. It lays its eggs on the surface of stagnant water, and hence low, marshy places are infested with *gnats*. The only place where the word occurs in the Bible is Matt. 23. 24, where our Saviour doubtless quotes a proverb well known to the Jews. The old English version has it "strain at a gnat," which is manifestly wrong, since it refers to the custom of straining wine, &c. in compliance with Lev. 11. 23, to avoid as food all *"flying"* and *"creeping* things,"* except the locust. The reader will thus see the appropriateness of the revised version—"strain out the gnat, and swallow a camel," *i. e.*, strain out or filter small insects, while the camel goes through. The Saviour applies it to such as scruple at trifling faults, and yet themselves commit the greatest sins.

7. APTERA. Insects without wings, such as *flea*, *spider*, *scorpion*, &c. Of *apterous* insects several are mentioned in the Scriptures.

FLEA (*Pulex Irritans*). The only place where it is mentioned in Scripture, is 1 Sam. 24. 14, and 26. 20, where David says to his persecutor, "Saul, after whom is the king of Israel come out? after whom dost thou pursue? after a *flea?*" "The king of Israel is come out to seek a *flea!*"

In this address to the king, David shows Saul how unworthy his royal dignity, to pursue after him, unworthy and insignificant, as he felt himself to be without any just reason whatever. The *flea* is noted for its extraordinary agility in jumping, it being capable of springing over 200 times its own length. It is generally associated with uncleanliness, and is a source of great annoyance to travelers vising Eastern countries.

In the order of apterous insects must also be placed the *"lice"* mentioned in connection with the third great plague brought upon Egypt (See Ex. 8. 16, 17; Psa. 105. 31). So many insects come under the general appellation of *"lice,"* that it is simply impossible to determine what insect it was. The Septuagint has *Sknipes* (flea), which rendering is followed by St. Jerome in the Vulgate. Dr. Adam Clark (notes on Ex. 8. 16) endeavors to show that it was the troublesome little insect known as the *Tick*, (*Acarus Sanguisugus*).

SCORPION (*Pedipalpi*). This crab-like creature is classed with the spider family (*Arachnida*), and may be briefly described as an insect allied to the spider. Its general appearance is that of a miniature lobster. In temperate climates it is from one to two inches in length, but in tropical countries it is much larger, sometimes exceeding ten inches. It is armed with a pungent sting, which is very poisonous and sometimes proves fatal (Rev. 9. 3–6). They abounded in the wilderness (Deut. 8. 15). In 1 Kings 12. 10, and 2 Chron. 10. 11–14, the word is evidently used to denote some instrument of punishment.

It is said that when in danger, from which it sees no means of escape, it stings itself to death. It draws itself together into an oval form, somewhat resembling an egg, and hence the Saviour's allusion in Luke 11. 11, 12. Scorpions are mentioned figuratively in Ez. 2. 6 and Rev. 9. 3–10.

SPIDER (*Aranea*). Spiders belong to the class *Arachinda*, which embraces many genera and hundreds of species, and are found in all parts of the world. Spiders may be divided according to their habits into five classes.

1. *Hunting Spiders*. They do not weave large webs, but make small tubes into which they live, and from which they sally forth and seize their prey. Some species are large, and can leap quite far. 2. *Wandering Spiders*. This kind do not weave webs, but throw out threads to capture their prey. 3. *Prowling Spiders*, which weave webs, and prowl about until

their prey is entrapped in it. 4. *Sedentary Spiders*, which spin the web, and then quietly wait in the middle, or edge of it, until the prey is entangled, when they seize it. 5. *Water Spiders*, which spin their webs on water plants. Spiders are carniverous, and eat almost all kinds of insects. In the tropics, spiders attain a large size. One species (*Mygale Avicularia*) is over two inches in length, and not only entraps large insects in its strong web, but also small birds. The spider is a very cunning insect. Its senses are very acute, especially sight. Most species have eight eyes. The bite of some is poisonous and painful, and in some instances proves fatal. Many species of spiders are found in Palestine. It is mentioned in Job 8. 14, " Whose trust shall be a spider's web." Isa. 59. 5, "And weave the spider's web," and Prov. 30. 28, " The spider taketh hold with her hands, and is in king's palaces."

HORSELEECH (*Hæmopis Sanguisorbia*) (Prov. 30. 24). Leeches belong to the order *Suctoria*, and genus *Anelida*. The leech is a kind of flattened worm. Some are aquatic, and others are land leeches. They are provided with sucking discs, in which are set several minute teeth, with which they cut the skin of other creatures and then gorge themselves with blood. The *medical leech* (*Hirundo Medicinalis*) is extensively used for blood-sucking in certain diseases. The "Horseleech" is much larger than the medical leech. They are very apt to enter the nostrils of horses and cattle while they are drinking; and when they have once taken hold of the nasal membrane, they hold so tenaciously that they sometimes cannot be extricated without tearing them asunder.

MOLLUSCA.

Although shells of various kinds abound in Palestine, there is no mention made of any, but the *snail* (*Helix*). In Lev. 11. 30, the snail is classed with unclean animals. Of the wicked, the Psalmist says, "As a *snail* which melted, so let every one of them pass away" (Psa. 58. 8). As there are many species of snails, differing much in size and color, nothing of a specific character can be said of the snail of the Bible. The reader is doubtless familiar with the snail, so that a description is unnecessary. Naturalists have described over 1400 species of snails, besides those found in a fossil state, buried in rocks and earth. Several species of snails are edible. Only one kind, however, (*Helix Pomotia*) is eaten by Europeans. This species is about two inches in length, and about the same in height. It is considered a delicacy in some parts of Europe.

PEARLS are several times mentioned in the Bible. On account of its value it is put in contrast with wisdom. "No mention shall be made of coral or of pearls, for the price of wisdom is above rubies" (Job 28. 18). The Saviour says, "The kingdom of heaven is like unto a merchantman, seeking goodly *pearls*, who, when he had found one pearl of great price, went and sold all he had, and bought it" (Matt. 13. 45, 46). He doubtless alludes to himself as that "pearl." It was of "great price" because of the great price paid by infinite love to make it available to man. Who can estimate the value of Christ's sacrifices? His humiliation, sufferings, death, &c. Pearls were used as ornaments (1 Tim. 2. 9; Rev. 17. 4). Very costly (Rev. 18. 12-16). The twelve gates of the heavenly Jerusalem are twelve pearls (Rev. 21. 12), (See Matt. 7. 6). The natural history of pearls is involved in doubt. They are small, oval concretions of great hardness and brilliancy, found inside of certain mollusks, chiefly in the *pearl oyster* (*Avicula Margaritifera*).

The general supposition is that they are the result of some injury to the creature, such as a grain of sand or other foreign substance becoming lodged within the shell. Nature is supposed to remove the cause of irritation by converting it into a pearl, which forms the nucleus, around which successive layers are deposited. This theory was verified by the great Swedish naturalist, Linnæus, who drilled minute holes into mollusks, and then inserted a grain of sand into the holes thus drilled. The result was the formation of pearl. For this discovery he was rewarded with a gift of £450 ($2,250) by his government.

Pearls differ much in size and color. Some are exceedingly valuable. The magnificent pearl which Cleopatra, the licentious queen of Egypt, dissolved, and then swallowed, was valued at 80,729 pounds, (over four hundred thousand dollars).

The word "*Purple*" occurs frequently in the Bible. Sometimes it means cloth of that color, as in Luke 16. 19, "Rich man clothed in *purple*." "Lydia, a seller of purple," &c., (Acts 16, 14). But it more frequently refers to the color itself (Jud. 8. 26; 2 Chron. 2. 7; Esth. 8. 15; Mark 15. 17), and many other places. Although the manufacture of this color is nowhere alluded to in the inspired portion of the Bible, but the phrase "purple of the sea" in 1 Macc. 4. 23 (*Apocrypha*) gives us some light. It is well known that the ancients obtained very fine and indelible purple and blue dyes from several kinds of marine mollusks. The Phœnicians greatly excelled in the manufacture of dyes, and it is probable that the Jews obtained their dyes chiefly from them. Purple was made from the shell-fish now known as *Murex Trunculus*, and blue from *Helix Ianthina*. (Linn.) The phrase in Ez. 27. 7, "Blue and purple from the isles of Elisha" is supposed to refer to Elis, a division of ancient Greece, noted for the commodities named.

ONYCHA, which occurs in Ex. 30. 34, as one of the ingredients of the sacred perfume, is supposed to have been obtained from a species of mollusk of the *Strombus* family, the horny operculum of which is said to be odoriferous.

CORAL (*Corallium*). Coral is the calcareous secretion of very minute insects of the class *Authrozoa*. The stone-like secretion is deposited in almost every imaginable shape. There are many species, all of which are governed by their own peculiar mode of coral building. Some are round, oval, branched. Some are shaped like a fan, cup, mush-room, &c. Several species are "reef builders." Their work is in great masses, sometimes hundreds of miles in length, and of unknown thickness. Large islands are formed by them in the seas, which in course of time acquire a soil from the wash of the sea. The birds and waves bring seeds from distant lands, and the result is a luxuriant growth of vegetation. Red and black coral are the most valuable kinds. Both kinds are found in the Mediterranean and Red seas. Red coral (*Corallum Rubrum*) is the most abundant, and is an important article of commerce in the East. It takes a very fine polish, and is used for ornamental purposes.

Mention is made of coral in the Old Testament, where on account of its preciousness it is put in contrast with wisdom. "No mention shall be made of *Coral* * * * for the price of wisdom is above rubies" (Job 28. 18).

Also Ez. 27. 16, "Syria was thy merchant by reason of the multitude of the wares of thy making. They occupied in thy fairs with emeralds, purple, and broidered work, and fine linen, and *coral* and agate."

MINERALS.

The ancients were familiar with nearly all the useful metals known at the present day. This fact implies that the arts attained a high degree of perfection in the early ages of man·kind. The general distribution of not only metals, but also of many kinds of rare and precious stones, shows conclusively that an extensive commerce existed between the nations of antiquity. Although Palestine, the home of the Jewish people, cannot be said to abound in native metals and precious stones, yet it is remarkable how plentifully they were supplied. Their chief source of supply probably was Phœnicia, the greatest commercial nation of antiquity. No other nation could compare with the Phœnicians in the art of ship-building and navigation, and Tyrian and Sidonian merchant vessels ventured to the uttermost parts of the earth, as far as then known, and brought the choicest treasures of the world to these cities, from which they were distributed, or in more modern phrase, wholesaled to surrounding nations. With this enter-prising nation the Jews were generally on amicable terms, and being contiguously situated to Palestine, both nations were mutually benefitted. A confederacy was formed between Solomon, king of the Jews, and Hiram, the Phœnician king, which was of great advantage to the Jews, and enabled Solomon to build the grandest Temple the world has ever beheld (1 Kings 5). With this same Hiram, king David had been on the most intimate terms of friendship (2 Sam. 5. 11 ; 1 Chron. 14. 1). So strong was the friendship between kings Hiram and Solomon, that they formed what might now be

termed an "International Trading Company" (2 Chron. 8.
17, 18). The object of these maritime expeditions was to enrich
their respective kingdoms with the products of distant lands,
such as the unknown Ophir and Tharshish (1 Kings 10. 22).
These facts are mentioned to throw some light on the probable
source of the enormous quantity of *silver* and *gold* and many
other precious commodities possessed by the Jews.

Our first inquiry will be directed to passages referring to
metals and metal working. It is a remarkable circumstance
that all the metals mentioned in the Bible, with one exception
(*antimony*), are mentioned in a single passage as follows:
"This is the ordinance of the law, which the Lord commanded
Moses, only the *gold* and the *silver*, the *brass*, the *iron*, the
tin and the *lead*, everything that may abide the fire, yea, shall
make it go through," &c. (Num. 31. 22).

GOLD. This is universally regarded as the most precious of
all metals, and has been used as a standard of value from
remote antiquity. It is the first metal mentioned in the Bible
(Gen. 2. 11), and its use is contemporaneous with man.

The Jews seem always to have had a plentiful supply of gold.
Abraham was rich in gold (Gen. 13. 2). Almost fabulous
quantities were used in the construction of the great Temple.
David had collected for this purpose one hundred thousand
talents (1 Chron. 22. 14). To this enormous amount king
Solomon added three thousand talents on his own account
(1 Chron. 29. 3), while the people added five thousand talents
more (1 Chron. 29. 7). If the reader will take the trouble to
reduce the amount to American money, he will find the sum
simply astonishing; and he will be enabled to understand the
assertion that "the king made *silver* and *gold* at Jerusalem as
plenteous as *stones*" (2 Chron. 1. 15).

Not only were large quantities of gold obtained from the
Phœnicians, as already stated, but the conquest and utter
destruction of the seven nations must have yielded to the Jews
a vast amount (Num. 31. 52; Judges 8. 26; 2 Sam. 8. 7;
2 Sam. 12. 30). Solomon received as a present from Hiram,
king of Tyre, one hundred and twenty talents of gold (1 Kings
9. 14), and a like amount from the queen of Sheba (1 Kings

10. 10 ; 2 Chron. 9. 9); we need not, therefore, wonder that his
household vessels and utensils were made of this precious
metal (1 Kings 10. 21). Because it is the most valuable of all
metals it is employed figuratively. Like gold we are purified
(Job 23. 10). A knowledge of God more desirable than gold
(Psa. 19. 10 ; Prov. 14. 16). Religious declension, "How is
the gold become dim " (Sam. 4. 1.) The trial of our faith
more precious than gold (1 Pet. 1. 7). Like pure religion, "I
counsel thee to buy of me gold tried in the fire, that thou
mayest be rich " (Rev. 3. 18). Words fitly spoken like "apples
of gold " (Prov. 25. 2). The streets of the Heavenly Jerusalem paved with gold (Rev. 21. 18). Gold is referred to more
than two hundred times in the Bible.

SILVER. This metal is first mentioned in Gen. 13. 2, where
we are told that Abraham was rich in cattle, *silver* and gold.
What was said in the preceeding article concerning the
sources whence the Jews obtained their gold, applies also to
silver. The two metals are closely associated, and very often
mentioned in connection with each other. David had provided
for the construction of the Temple "a thousand thousand
(one million) talents of silver (1 Chron. 22. 14). To which
Solomon added seven thousand (1. Chron. 29. 7), and the people ten thousand (verse 7). So plentiful was this metal during
the reign of Solomon, that it was no longer regarded as precious, while the vessels and utensils of the king's palace were
made of gold. "None were of silver: it was not anything
accounted of in the days of Solomon" (2 Chron. 9. 20). Like
gold it was also a medium of exchange (Gen. 23. 15). It was
extensively used for all manner of ornamental purposes. Its
process of manufacture applied to the experience of the righteous (Zech. 13. 9 ; Mal. 3. 3).

BRASS, TIN, ETC. The reader is doubtless aware that brass is
not a metal, but an alloy (a combination of several metals); as
now manufactured, brass is composed of copper and zinc. It
is considered doubtful whether the brass of the ancients was
identical with that now made, as zinc was not known as a distinct metal, until Paracelsus produced it in metal form in the
sixteenth century

Against these objections, however, it may be said that the addition of metalic zinc to copper is not necessary to con. stitute brass, since a fusion of *Calamine* (one of the ores of zinc) with copper ores, produces an alloy almost the same as modern brass. This view is sustained by the analysis of Roman coins, struck about the beginning of the Christian Era.

COPPER, which is the chief alloy of brass, is only mentioned in Ezra. 8. 27 : " Two vessels of fine copper, precious as gold." Of this it can only be said that copper as a pure metal cannot be intended, as it has always been quite abundant, and very much less in value than gold. Luther is probably right in rendering it *eherne* (brass), not that known to us as such, but an *alloy*, of which gold was the principal metal. This is the view held by most authorities, and there seems to be no doubt that the ancients made an alloy of copper and gold, which was of great value and beauty.

The "bright brass" of 1 Kings 7. 45, and "polished brass" (Dan. 10. 6), probably refers to this alloy. The only distinct reference, therefore, to copper is (2 Tim. 4. 14), "Alexander the copper-smith."

Since brass is mentioned in the Scriptures so frequently, and seems to have been so plentiful, and since its principle composition is copper, it seems somewhat remarkable that this metal (copper) is not mentioned distinctly (save Ezra 8. 27) in the Bible, and naturally leads to the supposition that in many instances we should read "copper" instead of "brass," and this applies especially to Deut. 8. 9, "A land whose stones are iron, and out of whose hills thou mayest dig brass" (copper ore). Also Job 28. 2, "Brass is molten out of the stone." The discovery of copper was coëval with iron, since Tubal-Cain, the father of metallurgy, worked both metals (Gen. 4. 22).

BRONZE. Bronze is an alloy of copper and tin. Although it is nowhere mentioned in the English version, it is quite probable that in some instances this alloy is intended, where the rendering is "brass." It is of fine grain and reddish yellow in color, and its manufacture is almost coëval with the human race.

Since it was in common use among all the enlightened nations of antiquity, it follows that the Jews were conversant with its manufacture; and especially so, since for many purposes it is preferable to brass or iron.

TIN. This metal is first mentioned in Num. 31. 22, as having been taken from the Midianites. Also (Isa. 1. 22; Ez. 20. 18, and Zech. 4. 10), "*Plummet,*" literally "*stone of tin,*" and Ezek. 27. 12, where we are informed that it was obtained in that far-famed country Tarshish, the location of which has for ages puzzled the learned. Since the tin is a very soft metal, it could not have been of great value to the ancients, except as an alloy of copper. In this form it is one of the most useful metals known. Combined with copper it forms "*Bronze,*" always rendered "brass" in the English version. Bronze can be made very hard, and some of the finest instruments and utensils are made of it. It takes a much finer polish than brass, and the "mirrors" and "looking-glasses" of the ancients were made of it (Ex. 38. 8). Also, "A molten looking-glass (Job 37. 18). This is the identical composition still used in the manufacture of the *Speculum* for refracting telescopes. The proportion for good reflecting surfaces is two parts copper to one of tin; for bells, cannon, tools, &c., the proportions vary.

LEAD. This metal, which serves so many useful purposes in the present age, does not seem to have been extensively employed by the ancients. Its various modes of application are comparatively modern inventions. The process of refining precious metals with quicksilver was not known to the ancients, and lead was used for the purpose. It is to this use of lead that the Prophet Jeremiah refers, "In their fire the lead is consumed, the smelting is in vain, for the evil is not consumed" (Jer. 6. 29). The Prophet Ezekiel (Ezek. 22. 18–22) speaks of lead in the process of metal refining, and gives it a forcible spiritual signification (see also Mal. 3. 2, 3). On account of its softness, this metal was made into plates or tablets, upon which to write (with a hard metal stylus). Such plates are often found in ancient ruins. To this Job evidently refers when he desires his words engraven "with an iron pen and lead" (Job 18. 24). The sense is obscure in the English, but finely brought out in the German version, thus, "*mit einem eisernen Griffel auf Blei,*" ("with an iron graver upon lead."),

On account of its weight, it is mentioned metaphorically, when the hosts of Pharaoh are said to have "sunk like lead" in the Red sea (Ex. 15. 10). The *plummet* used for sea soundings (Amos. 7. 7, 8; Acts 29. 28) was made of lead.

IRON. This is the most useful metal known to man for general purposes, and its use is coëval with the human race. The first iron founder seems to have been *Tubal Cain* (Gen. 4. 23). By a wise foresight into the wants of mankind, the all-wise Creator has so arranged it that the most necessary metal should also be the most plentiful and easiest of access. A metal so well known to the reader needs no further notice in this work. In the Old Testament it is mentioned very often. In the New Testament we read of an iron gate (Acts 12. 10). An iron surgical instrument (1 Tim. 4. 2). Breastplates (Rev. 9. 9).

STEEL. This metal is refined iron, charged with carbon. It is considered doubtful whether the ancient Hebrews were acquainted with steel. The only passage in the Bible where steel is referred to, is probably Jer. 15. 12, "Shall iron break the northern iron and the steel?" It is generally supposed that the allusion is to the product of the Chalybes of the Pontus, who were the most famous iron-workers of antiquity. "Steel" also occurs in 2 Sam. 22. 35; Job 20. 24; Psa. 18. 34, but is rendered "brass" by Luther and most other translators. The "bow of steel," mentioned in 1 Sam. 22. 35, was probably brass (so Luther). This metal when properly tempered is almost as flexible as steel.

ANTIMONY. Although this metal is nowhere mentioned in the English version, it is obviously implied in the number of passages. Antimony is a fine grained, but brittle metal, of a bluish-white color. It is very useful as an alloy of other metals. Antimony was not known to the ancients as a metal, but in its unreduced state it was used as a paint. This paint or dye was well known to the Jews, and occurs in the Bible under the Hebrew word *puch*. This word (puch) is only indirectly translated as follows: In 2 Kings 9. 30, the queen Jazebel is said to have "painted her face." Here the word puch occurs as the pigment. The same word occurs in Jer. 4. 30, "Though thou rentest thy face with painting," (puch). Also

(Ezek. 23. 40) "Paintedst thine eyes." Literally "puched thine eyes." The ores of lead and antimony were mixed with oil, and applied to the eye-brows and eye-lashes, to improve the appearance. This same word also occurs in 1 Chron. 29. 2, "Glistering stones," literally "stones of puch." Also the beautiful prophecy of Isaiah (Isa. 54. 11), concerning the prosperity of Messiah's reign, "I will lay thy stones with fair colors," (literally with puch).

MINING and METAL WORKING. There are various allusions in the Bible to the process of obtaining minerals, and preparing the metal for use. For a graphic description of mining see Job. 28. 1–11.

The references to metal working are as follows: Casting (Ex. 25. 12–16; 2 Chron. 4. 17; Isa. 40. 19). Welding (Isa. 51. 7). Hammering into sheets (Num. 16. 38; Isa. 44. 12; Jer. 10. 4–9). Overlaying (Ex. 25. 11–24; 1 Kings 6. 20; 2. Chron. 3. 5; Isa. 40. 19; Zech. 13. 9). Refining (Psa. 11. 6; Prov 17. 3, &c.; Isa. 1. 25; Jer. 6. 29; Mal. 3. 2, 3.)

GEMS AND PRECIOUS STONES.

———

A love for the beautiful in nature is inherent in man, and early led him to seek the chrystallized forms of matter, commonly called *Gems*. In the Scriptures precious stones are early recognized as valuable and desirable objects, and are mentioned in connection with gold, as products of *Havilah.* "And the gold of that land is good, there is *Bdellium* and the *Onyx stone*" (Gen. 2. 12). Mankind early attained considerable proficiency in the art of Gem cutting. Some very ancient specimens recovered from Assyrian ruins are hardly surpassed by modern productions, while the Etruscan and Greek specimens still remain unrivalled. The number of precious stones known to the ancients was comparatively large, and modern crystallography has added but very few to the list of gems.

Nearly all the precious stones mentioned in the Bible occur in three different passages. The first is the enumeration of the stones, twelve in number, in the Breastplate of the High Priest (Ex. 28. 17; 39. 10). Among the ornaments of the king of Tyre there are nine gems enumerated (Ezek. 28. 13).

In the graphic description of the Heavenly City (Rev. 21. 18, &c.), its foundations are said to be of precious stones, twelve in number, almost similiar to the gems of the High Priest's breastplate. Besides the gems enumerated in the above lists, a number of others occur in the Bible. The identity of the various gems of the Bible is, perhaps, impossible in all cases. The renderings of the English version are not all uniform with other translations, but in the uncertainty which attends the Hebrew originals, it is believed that the renderings given in the English are as trustworthy as that of any other version.

ADAMANT. The Hebrew original is *Shamir*, and occurs as follows: (Jer. 17. 1), "With the point of a *Diamond;*" (Ezek. 3. 9), "As *adamant* harder than *flint;*" (Zech. 7. 12), "They made their hearts as an *adamant* stone."

The Vulgate makes it the diamond in all the passages, and is also that given by Luther. With few exceptions it is agreed that the diamond is meant.

Adamant is a general term for *Corundum* or *Adamantine. Spar*, which next to the diamond is the hardest substance known. The finer forms of corundum are known as gems under various names, according to color, not differing substantially, except in the coloring matter. The diamond derives its name from adamant, but is an essentially different substance. While adamant (*Corundum*) is essentially *Alumina*, the diamond is pure carbon, and with the application of the blowpipe can be burned without leaving any residium.

Diamonds are generally colorless and clear as water. They are, however, found in almost every color, as the result of impurities. They are very brilliant, highly electric, and when slighly rubbed, emit a faint light in the dark.

Diamonds are obtained in various parts of the world, chiefly in India (*Golconda*), and Brazil. The largest diamonds are almost beyond price. The "Great *Koh-i-noor*," now the property of Queen Victoria, was obtained from India, upon the surrender of the Punjaub to England. Its weight was 900 carats, but by repeated cuttings has been reduced to 123 carats. It has been valued at 120,664 pounds sterling, ($603, 220).

The finest diamond, perhaps, now known, belongs to the sovereign of Mattan. For this fine stone he refused an offer of $500,000, two war brigs, fully equipped, a number of cannon, and a large quantity of ammunition, which offer was made to him by the governor of Borneo.

AMETHYST. This stone was the ninth in the High Priest's breastplate (Ex. 28. 19; 39. 12), and the twelfth in the foundation of the New Jerusalem (Rev. 21. 20). The amethyst

8

of the Bible (Oriental Amethyst) is a rare and beautiful
variety of *corundum.* It is a transparent stone of a strong
blue, or deep red color, although other colors sometimes
prevail. The stone commonly known as amethyst, of a violet
blue, or purple color, is merely *Amethystine quartz,* entirely
different from Oriental Amethyst.

AGATE-ACHATES. The eighth stone in the High Priest's
breastplate (Ex. 28. 19 ; 39. 12). Agate derives its name from
the River Achates in Cicily (now called the Drillo), where
many were obtained in ancient times. Agate is a variegated *Chalcedonic quartz.* Some kinds are composed of many
strata, the lines arranged in *zig-zag* order. Such kinds are
called "*fortification agates.*" Sometimes it is found of a semi-
transparent, milky-white, with figures resembling patches of
moss or loose hair, such are called "*moss agates.*"

BDELLIUM. Mentioned in Gen. 2. 12. As now known, this is
a resinous gum obtained from several different Oriental trees.
Its occurrence in the Bible is attended with much uncertainty.
Some eminent authorities maintain that the original in the
above passage refers to some kind of stone.

BERYL. (Hebrew, *Shoham*). This was the tenth stone in the
breastplate (also Dan. 10. 6, and Rev. 21. 20.) This same
word (*Shoham*) occurs in other places, but is differently
rendered. The mineral character of the Beryl is as follows :
Specific gravity, 2.7 ; Silica, 66 per cent. ; Alumina, 17 per cent. ;
Glucina, 15 per cent. ; Iron Oxide, 2 per cent. It is greenish
in color, and the very clear, sea green beryls are called *Aqua
marines,* and are quite valuable.

CARBUNCLE. The third gem in the breastplate (Ex. 28. 17;
39. 10). (See also Isa. 54. 12, Ezek. 28. 13.) It is generally
agreed that the Carbuncle is the stone now known as the
Garnet. There are several varieties of this stone, the finest is
"precious garnet," also known as *Almandine.* It is a very
fine stone of a deep crimson-red color.

CHALCEDONY (Rev. 21. 19). A stone in the foundation of the
New Jerusalem. This stone derives its name from *Chalcedon*
in Bithynia. Chalcedony is a variety of quartz. It is found
massive. It is a semi-transparent, lac lustre stone, generally

milk-white, or bluish white, although other colors are not uncommon.

CHRYSOPRASUS. The name of the stone in the tenth foundation of the New Jerusalem (Rev. 21. 20). This is essentially a variety of Chalcedony, of a deep apple green color, caused by the presence of oxide of nickel.

CHRYSOLITE. The seventh foundation of the Heavenly City (Rev. 21. 20). This is a transparent of a fine green color, and possesses the property of double refraction.

Its constituents are *silica, magnesia,* and *protoxide of iron.*

DIAMOND (Ex. 28. 18; 39. 11; Jer. 17. 1). The most precious of all gems. Described under "Adamant."

EMERALD. The fourth stone in the breastplate (Ex. 28. 18; 39. 11). In the fourth foundation of the New Jerusalem (Rev. 21. 19). (See also Ezek. 27. 16, 28. 13; Rev. 4. 3). This fine stone is substantially the same as the *Beryl,* except in its iron pigment, which is 2 per cent. *chromic iron,* instead of 2 per cent. oxide of iron as in the Beryl. It is of a rich green color, and of a vitreous lustre.

JACINTH (Rev. 9. 17; 21. 20). The identity of this stone is by no means certain. In the first passage Luther renders it merely " *Gelbe,*" (yellow). In the second "*Hyacinthe.*" The equivalent of Jacinth.

The Hyacinth is almost a pure *Zircon.* Its mineral character is as follow: Specific gravity, 4.6; Zirconia, 67 per cent.; Silica, 33 per cent. It is a clear and beautiful stone and of adamantine lustre. Color generally red. By some authorities this stone is supposed to be the *Oriental Sapphire,* which is an altogether different gem.

JASPER. One of the gems in the High Priest's breastplate (Ex. 28. 20; 39. 13). Also one of the foundations of the New Jerusalem (Rev. 21. 18, 19). (Also Ezek. 28. 13; Rev. 4. 3; 21. 11).

Jasper is regarded by mineraloyists as one of the forms of Quartz. It is an opaque stone of considerable hardness, and takes a magnficient polish. The color of Jasper is not uniform, they are found of many hues; some are entirely of one color, while others are variegated.

ONYX. A stone in the fourth row of gems in the breastplate of the High Priest (Ex. 28. 20; 39. 13). Also mentioned in Job 28. 16, and Ezek. 28. 13). Onyx is a variety of agate; black and white, or dark brown and white stripes of chalcedony alternate. In some rare specimens a third colored band occurs. It appears, however, that in ancient times the name was not always confined to the above variety.

LIGURE. A stone in the third row of gems in the breastplate (Ex. 28. 19; 39. 12). The identity of this stone is involved in doubt. Authorities differ very much, and the researches of critics have failed to establish its identity. It is improbable that the stone known to mineralogists as *Ligurite*, is intended, although so held by some. *Opal* and *Amber* have been proposed, as also some other stones. Where so much uncertainty prevails, no additional light can be obtained by a further investigation.

RUBY (Job 28. 18; Prov. 8. 11; 3. 15; 20. 15; 31. 10; Sam. 4. 7). The Ruby is a variety of *Sapphire*, and in value is next the diamond. Its color is red in its various shades.

SAPPHIRE. A gem in the second row of the breastplate (Ex. 28. 18; 39. 11), and also in the foundation of the New Jerusalem (Rev. 21. 19). (Also Ex. 24. 10; Job 28. 16; Lam. 4. 7; Ezek. 1. 26; Cant. 5. 14, and Isa. 54. 11). The *Sapphire* is the purest kind of *Corundum*, with a small addition of *Chromic oxide*, which determines its color. In the absence of the oxide it is perfectly transparent and highly brilliant, and cannot be distinguished from a diamond, except by an expert. Next the diamond it is the finest and costliest gem.

The rarest variety of Sapphire is the red (Ruby), already mentioned; another lovely variety is the blue, called by the ancients "*Lapis Lazuli.*" Of this kind the celebrated paint called "*Ultra marine*" is made.

SARD-SARDIUS. A stone in the first row of the breastplate (Ex. 28. 17; 39. 10) and the sixth foundation of the New Jerusalem (Rev. 21. 20). (Also Ezek. 28. 13). This is a quartz gem of considerable fineness. It is the deep blood-red *Cornelian* of the present day. It derives its name from Sardis in Asia Minor.

SARDONYX. The fifth foundation in the New Jerusalem (Rev. 21. 20). It is a *Sard-onyx*, that is, an *onyx* containing bands of *sard*.

TOPAZ. One of the gems in the breastplate (Ex. 28. 17 ; 39. 10), and the ninth foundation of the New Jerusalem (Rev. 21. 20). (Also Job 28. 19 ; Ezek. 28. 13). The topaz is one of the varieties of *Sapphire*. Its general color is golden yellow, although other shades of yellow are found; from the darker shades of this color it passes into red. It is then called a Ruby.

AMBER. The Hebrew word *Chashmal*, rendered "Amber" in the English version, occurs in Ezek. 1. 4–28, and 8. 2). Considerable uncertainty attends the true meaning of the word. Some European versions do not translate it "Amber color," but merely give the equivalent of *Brilliancy*—(So the German, which renders it *"lichthelle."*) The English translators probably followed the *Septuagint* and *Vulgate*, both of which give *Electrum*, which is the equivalent of *Amber*, and also of a brilliant metal composed of gold and silver.

Amber is a hard, resin-like, semi-transparent substance, quite glossy, and is much used for beads and ornaments. It occurs as a mineral, and is now mostly obtained along the shores of the Baltic sea. From the fact that petrified leaves and insects are found in Amber, it is supposed to be the fossil resin, or gum, of some species of coniferous tree, now extinct. Amber is highly electric when rubbed.

NON-METALLIC SUBSTANCES.

ALABASTER. This substance occurs only in Matt. 26. 7, and Mark 14. 3, where it is said that a woman brake an *alabaster box*, containing a very precious ointment of spikenard, and with it anointed the head of our Saviour, as he was eating in Bethany, in the house of Simon the leper. Alabaster is a mineral substance, and occurs as rock in many places. There are two kinds, similar only in name, as they differ in their chemical composition very much. *Soft alabaster* is a white and semi-transparent variety of *Gypsum* (sulphate of lime). It is abundant in Italy, and is manufactured into vessels and ornaments. The other kind is composed of carbonate of lime, and is of a more crystalline nature. It is this variety that was used in the manufacture of perfume bottles or vessels.

It was quite abundant in Egypt near a town called *Alabastron*, where the manufacture of Alabaster vessels was extensively carried on. It derived its name from the place where obtained.

It is probable that instead of destroying the vessel, Mary broke the seal, so that the contents could be used. It is said to have been a pound in weight, which was Troy weight (12 oz.). The value of the ointment, according to Judas, was 300 pence, or $40 in American money.

CHALK. Chalk is a soft, white rock, composed of carbonate of lime, and is, therefore, chemically identical with limestone. Various kinds of limestone are met with in Palestine. Three different groups of limestone rocks have been traced by the Palestine exploration society. Some of the most interesting topographical features of Palestine are the result of the calcareous nature of its rock beds, such as irregularity of surface, abrupt precipices, numerous caves, &c., caused by the wearing away, or decomposition of the limestone rock. Chalk only occurs once in the Bible (Isa. 27. 9).

CLAY. This substance is frequently mentioned as the material of which bricks, pottery and seals were made. Clay is a mineral substance possessing considerable plasticity and tenacity qualities, which make it very useful to man.

Clay is produced by the decomposition of other minerals. Its character is determined by the mineral which predominates in its composition. Hence we have various kinds of clays, such as *Kaolin, Pipe Clay, &c.*

The first mention in the Bible of the economic use of clay is the reference to Rebekah's pitcher (Gen. 24. 14). Gideon's little army hid their lamps in "pitchers" (Judges 7. 16–19). In Jer. 32. 14, is a reference to earthen safes.

Clay was, of course, used in the manufacture of *bricks.* Two methods of making them are indicated in the Scriptures. The Assyrian method was to mix the clay with "slime" (Asphaltum), and burn them in a kiln (Gen. 11. 3; Jer. 43. 9). In Egypt bricks were made by adding cut-straw, and drying them in the sun. Such bricks the Hebrews made during their bondage there (Ex. 1. 14; 5. 7). In ancient times clay was also used for seals (Job. 38. 14).

FLINT (Deut. 8. 15; Psa. 114. 8; Isa. 5. 28; Isa. 50. 7; Ezek. 3. 9). Flint is a stone intermediate between *quartz* and *opal.* It is a mineral of great hardness, and the readiness with which it strikes fire on hard metals is well known. It was used for this purpose for ages, until the invention of lucifer matches. It is found in many colors. It occurs chiefly in chalk formations.

MARBLE. Marble is simply the purest form of limestone, of a crystalline structure. It is snow-white when pure carbonate of lime. The various colored marbles, such as black, red, grey, blue, &c., owe their color to the presence of other substances, such as iron, manganese, &c. Since remote ages, marble has been used for building and ornamental purposes. David provided marble in abundance for the construction of the Temple (1 Chron. 29. 2). There were "pillars of marble" in the palace of king Ahasuerus (Esth. 1. 6). It is mentioned metaphorically, "His legs are as pillars of marble" (Sol. Song 5. 15).

NITRE. It is universally agreed that the mineral known to us as nitre (saltpetre) does not occur in the Bible. The rendering of *nitre* instead of *natron*, which is the substance referred to, arose from a misapprehension of the chemical difference between the two minerals.

Nitre (*Nitrate of potash*) and *Natron*, (*Sesquicarbonate of soda*) are very different substances. *Nitre* does not effervesce by the application of acid, but *natron* does, thus explaining (Prov. 25. 20), "As vinegar (acid) upon nitre (*natron*), so is he that singeth songs to a heavy heart." Having reference to the effervescence caused by the action of the acid upon soda. This mineral has the properties of common washing soda, and was chiefly used for cleansing purposes. Hence the Prophet's declaration, "Though thou wash thee with nitre (natron) * * * yet thine imquity is marked before me" (Jer. 2. 22). Both the Septuagint and Vulgate give natron as the substance intended, and the evidence, therefore, seems conclusive.

This mineral is chiefly found in Egypt, where it is deposited on the bottom of the Natron Lakes—(six in number). During the hot season the water is evaporated, and the natron is found thickly incrusted on the bottom. The chemical properties of natron are *carbonate, sulphate*, and *muriate of soda*.

SALT. It is evident that the use of salt is coëval with man. Both man and beast have a natural craving for it, which does not seem to be the result of mere habit. It is doubtful whether any other non-metalic mineral is of greater value to man than salt. Salt (chemically known as *sodium chloride*), is found in its natural state all over the world, either in great, solid masses like rocks, or held in solution by water, from which it is separated by evaporation.

The uses of salt are numerous. It is a very valuable remedy, and its medicinal qualities are not sufficiently made use of at the present day. Many of the qualities of salt are alluded to in the Bible. As a *condiment* for food, "Can that which is unsavory, be eaten without salt?" (Job 6. 6). Also for cattle

(Isa. 30. 24. Savory provender in the margin). Salt, when sparingly applied, is a stimulant to the soil, but when unduly applied destroys vegetation, hence it became emblematic of destruction (Judges 9. 45; Deut. 29. 23; Zeph. 2. 9). It is well known that salt loses its strength by exposure, and hence our Saviour's application Matt. 5. 13.

Its antiseptic and preserving qualities are several times alluded to in the Bible. In ancient times new-born infants were washed in *salt* water, and this custom is mentioned in Ezek. 16. 4.

SAND is formed by the disintegration of rocks. In some parts of the globe great tracts of country are entirely covered with sand. It is often mentioned in the Bible. It is an emblem of great numbers (Gen. 32. 12, &c.). Expressive of weight (Job 6. 3). Moses hid the victim of his vengeance in the sand (Ex. 2. 12). Sand banks on the sea shore referred to (Jer. 5. 22). The foolish man builds his house on the sand (Matt. 7. 26).

PITCH-SLIME (*Septuagint,* "*Asphaltos.*") These two terms refer to the substance now known as *Asphaltum.* It is a compact bituminous mineral, of a black, or dark brown color. It is found abundantly in the vicinity of the Dead Sea (Gen. 14. 10). From Scripture we learn that it was used to render vessels water-tight (Noah's ark, Gen. 6. 14). (Moses, Ex. 2. 3). Used in brickmaking (Gen. 11. 3).

According to the ancient historian Strabo, it was used by the Egyptians for fuel, and also for embalming the dead (Isa. 34. 9).

BRIMSTONE. This mineral, on account of its inflammability, is most frequently mentioned in the Bible as emblematic of the wrath of God (Gen. 19. 24; Deut. 29. 23; Psa. 11. 6; Ezek. 38. 2; Isa. 30. 33; Rev. 21. 8, &c.).

Brimstone (*Sulphur*) is abundantly found in the vicinity of the Dead Sea, and is probably the product of the numerous sulphur springs of that region.

✳ FINIS. ✳

INDEX.

www.ingramcontent.com/pod-product-compliance
Lightning Source LLC
Chambersburg PA
CBHW021939190326
41519CB00009B/1069